Deactivation of Catalysts

Deactivation of Catalysts

R. HUGHES

*Department of Chemical Engineering,
University of Salford, Salford, UK*

1984

ACADEMIC PRESS

(Harcourt Brace Jovanovich, Publishers)
London Orlando San Diego San Francisco New York
Toronto Montreal Sydney Tokyo São Paulo

ACADEMIC PRESS INC. (LONDON) LTD.
24–28 Oval Road
London NW1 7DX

U.S. Edition published by
ACADEMIC PRESS INC
(Harcourt Brace Jovanovich, Inc.)
Orlando, Florida 32887

Copyright © 1984 by
ACADEMIC PRESS INC. (LONDON) LTD.

All Rights Reserved

No part of this book may be reproduced in any form by photostat, microfilm, or any other means, without written permission from the publishers

British Library Cataloguing in Publication Data
Hughes, Ronald
 Deactivation of catalysts.
 1. Catalysis
 I. Title
 660.2'995   TP156.C35

 ISBN 0-12-360870-8
 LCCCN 83-71679

Typeset by Eta Services (Typesetters) Ltd., Beccles, Suffolk
Printed in Great Britain by St Edmundsbury Press, Bury St Edmunds, Suffolk

Preface

The use of catalysts is widespread for numerous important processes throughout the chemical and petroleum processing industries. Unfortunately, most catalysts suffer a loss of activity with time on stream and the life of even the most robust catalyst is limited. Loss of activity is frequently accompanied by the equally serious problem of loss of selectivity if complex reactions are being processed. This book is generally addressed to this problem which in practical terms involves the problem of determining how long a catalyst can remain in the reactor before replenishment or reactivation of the catalyst is necessary and the more fundamental problem of minimizing this loss of activity. Monitoring of catalyst activity is commonly done in industrial processes, for example by measuring the temperature profiles, but this is not always indicative of the mechanism of deactivation which is occurring.

The problem of catalyst deactivation has attracted increasing attention in recent years as is evidenced by the number of papers published in this area, and a greater understanding of the underlying deactivation processes is now possible. The purpose of this monograph is to bring together this information as a more cohesive whole.

The conventional divisions of catalyst activity loss in terms of thermal sintering, impurity poisoning and fouling are still generally convenient and have been adhered to in this text, although with continuing research these divisions are now less rigid than formerly. Essentially, the material of this book is confined for the most part to gas–solid heterogeneous catalytic systems although some three-phase systems are included such as trickle bed processing. Homogeneous systems are not considered.

The approach throughout has been in terms of chemical reaction engineering concepts, involving mathematical description of the phenomena where possible. It is the author's view that this is the best way of interpreting the sometimes complex phenomena arising from consideration of reactions subject to deactivation in both catalyst pellets and catalytic reactors.

Hence, in this book, after a general review of catalyst deactivation the effects of the various types of deactivation in single pellets are considered separately. The analysis is then extended to catalytic reactors and the effects of each kind of deactivation in these are considered together with the optimum operating policies for reactors subject to deactivation. Finally, the problems arising from catalyst regeneration are considered. This is, in itself, a very wide topic; the treatment of this important aspect has inevitably had to be limited to the main features of this subject.

The book should be of interest to chemical engineers or chemists working in industry or for those starting research on catalytic systems. It also contains material suitable for graduate courses in chemical engineering and for some final year undergraduate courses in chemical reaction engineering or reactor designs.

Inevitably in a work of this kind the publications and experience developed in the author's own group have been drawn upon to a fair extent. However, the important contributions made to this whole area by many notable authorities in this field are recognized and are cited in the text. The stimulation and encouragement received from the various individuals is gratefully acknowledged.

Particular thanks are due to the many colleagues who by their sustained efforts over the years helped to enlarge the author's interest in this field. Mention must be made of Dr H. P. Koh and Dr B. Valdman and especially of Dr Ezra Kam, Dr Andrea Byrne and Dr P. A. Ramachandran who contributed immensely to the outcome of such successful research.

A great debt of gratitude is owed to Mrs Audrey Saunders and Mrs Dorothy Scott who typed and retyped drafts of the manuscript, cheerfully and most ably.

JANUARY 1984   RONALD HUGHES

To Sylvia, Rosamund and Graham

Contents

Preface v

1 Introduction

1.1 General introduction. 1
1.2 Classification of catalyst deactivation processes 3
References 6

2 General Aspects of Catalyst Deactivation

2.1 Poisoning of catalysts 7
   2.1.1 Poisoning of metallic catalysts 9
   2.1.2 Poisoning of non-metallic catalysts 9
   2.1.3 Poisoning of bifunctional catalysts. 12
2.2 Coke formation on catalysts 13
2.3 Metal deposition on catalysts 19
2.4 Sintering of catalysts. 22
References 27

3 Diffusion and Deactivation of Catalysts

3.1 Introduction 29
3.2 Mass and heat transport in catalyst pellets 31
   3.2.1 External mass and heat transfer. 31
   3.2.2 Intrapellet mass transfer 33
   3.2.3 Intraparticle heat transport 38
3.3 Diffusion and reaction in porous catalysts 38
3.4 Deactivation and diffusion 43
   3.4.1 Analogy between selectivity and deactivation mechanisms 43
   3.4.2 Catalyst deactivation 43
3.5 Correlations for activity decay 49
3.6 Separable and non-separable kinetics 52
References 53

4 Catalyst Deactivation by Sintering

4.1 Introduction 55
4.2 Structural features of catalysts relevant to sintering 56
4.3 Experimental data 57
   4.3.1 Experimental methods 57
   4.3.2 Experimental results 58
4.4 Mechanisms of sintering for supported metal catalysts 63

4.4.1	Particle migration and coalescence.	63
4.4.2	Interparticle transport	64
4.4.3	The particle migration model	65
4.4.4	The atomic migration model	69
4.4.5	Comparison of models for sintering	72
4.4.6	Experimental work of Wynblatt and Gjostein	74
4.5	General summary of current sintering knowledge	77
References		79

5 Catalyst Deactivation by Poisoning

5.1	Some common poisoning processes	81
5.1.1	Steam reforming catalysts.	81
5.1.2	Low-temperature shift catalysts.	83
5.1.3	Methanation catalysts	84
5.1.4	Oxidation catalysts for formaldehyde production	84
5.1.5	Temporary (reversible) poisoning of ammonia catalysts.	84
5.1.6	Reversible poisoning of hydrogenation catalysts	85
5.2	Minimization of poisoning	85
5.3	Poisoning of automobile catalysts	87
5.4	Analysis of poisoning processes	88
5.4.1	Irreversible poisoning	88
5.4.2	Reversible poisoning	97
5.4.3	Non-isothermal effects	103
5.5	Effect of poisoning on catalyst selectivity	107
References		108

6 Catalyst Deactivation by Fouling

6.1	Fouling by coke deposition	110
6.1.1	Coke formation processes.	112
6.1.2	Experimental results from laboratory coking studies.	113
6.1.3	General approaches to developing a theory for deactivation by coking	116
6.1.4	Modelling of coking in single catalyst pellets	119
6.2	Catalyst deactivation due to deposition of impurities in the feed onto the catalyst	140
6.2.1	Modelling studies of pore plugging deactivation	140
References		143

7 Deactivation in Catalytic Reactors

7.1	Introduction	145
7.2	Poisoning in fixed bed reactors	147
7.2.1	Isothermal analysis.	147
7.2.2	Non-isothermal analysis	154
7.3	Coking of catalytic reactors	157
7.3.1	Isothermal coking in fixed bed reactors	159
7.3.2	Non-isothermal analysis of coking in fixed bed reactors	162
7.4	Thermal sintering of reactors	183
7.5	Reactor dynamics and catalyst deactivation	184
References		189

8 Optimization of Deactivating Reactor Systems

8.1	Comparison of various reactor types under deactivating conditions	190
8.2	Optimization of deactivating reactors	199
	8.2.1 Optimal temperature policies	200
	8.2.2 Other optimal policies	206
References		210

9 Regeneration of Deactivated Catalysts

9.1	Feasibility of regeneration	212
9.2	Description of coke deposit and kinetics of regeneration	214
9.3	Regeneration of fluidized bed catalysts	218
9.4	Regeneration of coked catalyst pellets	219
9.5	Regeneration of fixed beds containing coked catalysts	242
References		260
Subject Index		261

Notation

A	pre-exponential factor in Arrhenius expression; constant in equation (2.1)
A'	constant
A_c	cross sectional area of reactor
A_p	poisoning pre-exponential factor
a	dimensionless reactant concentration, C_A/C_{Ao}; stoichiometric factor
B	constant
B'	extent of reaction
b	dimensionless concentration of species B, C_B/C_{Bo}; constant
b_{ij}	rate of collision between particles comprising i and j units
C_A	concentration of reactant A
C_B	concentration of product B
$C_A{}^*, C_B{}^*$	dimensionless concentrations of A and B
C_c	concentration of coke or carbon
C_d	concentration of deactivating species
C_i, C_j	concentration of particles containing i and j units
C_k	number of crystallites per unit area of support
C_L	concentration of reactant at point L from pore entrance (Chapter 3)
C_o	concentration in bulk gas
C_0	concentration at mouth of pore
C_O	oxygen concentration in pellet
C_{Og}	concentration of oxygen in bulk gas
C_P	adsorbed poison concentration
C_s	concentration at surface
C_W	concentration of water vapour
C	dimensionless concentration of species C, C_C/C_{Co}
c_g	specific heat of gas
c_p	specific heat
c_s	specific heat of solid (catalyst)
D	dispersion (of metal crystallites); diffusivity in gas phase
D_{AB}	bulk diffusion coefficient for molecules A in binary mixture A–B
D_a	surface diffusivity of metal atoms; diffusivity in macropores
D_e	effective diffusivity
D_{eA}	effective diffusivity of species A
D_K	Knudsen diffusion coefficient (D_{KA} for species A)
D_{ep}	effective diffusivity of poison
D_i	diffusivity in micropores

Notation

D_{ij}	binary diffusivity of species i in j
D_{im}	diffusivity of species i in mixture m
D_p	particle surface diffusivity
d	order of deactivation; diameter of metal crystallite; diameter of reactor tube
d_p	diameter of particle
E	activation energy
E_d	activation energy for deactivating reaction
E_i^*	exponential integral term
F	feed rate
F_0	initial feed rate
F_1	parameter defined in equation (9.10)
g	ratio of facet diameter to particle radius (Chapter 4)
G	mass flow rate
G_i	growth of metal crystallite during sintering
ΔH	enthalpy of reaction
$(\Delta H)_m$	heat of atomization
$(\Delta H)_s$	atom–support interaction energy
h	heat transfer coefficient
h_{K_A}, h_{K_B}	heats of absorption for species A and B
J_H	dimensionless heat transfer coefficient
J_M	dimensionless mass transfer coefficient
j_D	j factor for mass transfer [equation (3.5)]
j_H	j factor for heat transfer [equation (3.5)]
K	constant; thermal conductivity
K'	constant
K_A, K_P	adsorption constants
K_A^*, K_B^*	dimensionless adsorption constants
K_e	effective thermal conductivity
K_i	equilibrium adsorption constant for species i
K_i^*	dimensionless adsorption constant for species i
K_{ij}	rate constant for crystallite containing i and j units
k	rate constant
k_i	rate constant for species i
k_B	Boltzmann constant
k_c	mass transfer coefficient
k_d	rate constant for deactivation reaction
k_m	mass transfer coefficient
k_o	rate constant at specified conditions
k_p	poisoning rate constant
k^o, k_i^o	rate constant for non-deactivated reaction
k_d^o	rate constant for deactivation under specified conditions
k_c, k_c^*	second order rate constant for reaction of carbon in coke based on pellet and bulk gas temperatures respectively
k_H, k_H^*	surface reaction rate constant for hydrogen based on pellet and bulk gas temperature respectively

Notation

L	length
L_1, L_2	atom loss for crystallites 1 and 2 in sintering
L_e	actual path length of gas molecules in porous solid
l	bed length; jump distance of atom
M	molecular weight
M_A	molecular weight of species A
M_B	molecular weight of species B
m	reaction order; mass of atom or molecule; exponent
$\bar{m}$	average molecular weight of gas mixture
N	number of pores
N_A	molar flux of species A
N_B	molar flux of species B
N_E	number of pores effective for demetallation
N_{AS}	superficial molar flux of species A
$N(r)$	number of pores of radius r
N_s	number of solid diffusion transfer units in bed
N_t	number of transfer units
n	constant
n_p	number of pores
P	pressure
p	dimensionless poison concentration, C_p/C_{po}
p_A, p_R	partial pressures of species A and R respectively
p_i	partial pressure of species i
Q	heat transferred; volumetric fraction of solid phase satured with poison
q_H	C_{Ho}/C_{og} in carbon/hydrogen model for regeneration of catalysts
q	fraction of pore surface activated
q_s	saturation concentration of poison in solid phase
R	gas constant; radius of pellet
$\bar{R}$	overall reaction rate
R^*	reaction rate
R_d^*	reaction rate under deactivating conditions
R_{ij}	collision radius between particles containing i and j units
r	radial distance in pellet; reaction rate
$\bar{r}$	mean particle radius
r_A	reaction rate for species A
r_B	reaction rate for species B
r_c	rate of coking
r_e	mean pore radius
r_p	radius of pore
S	activity of catalyst; metal area of supported metal catalyst; selectivity ratio
S_o	initial catalyst activity; initial metal area
S_b	specific surface area
S_F	space velocity of feed
S_g, S'	external surface area of catalyst pellet
S_k	exposed surface area of particle containing k units

$\bar{S}$	mean activity
T	dimensionless time, temperature
T_c	temperature of coolant
T_o	temperature in bulk gas
T_s	temperature at surface
T^*	dimensionless temperature
t	time
t^*	characteristic time
t_d	time on stream
t_m	optimum time
$\bar{t}$	ratio of time to deactivation time
U	uptake of gas on catalyst
u	gas velocity; parameter defined in equation (9.14)
V	metal molar volume
V_b	volume of catalyst pellet
V_g	porosity of pellet
v	velocity
X	conversion; dimensionless distance in flat slab
x	length dimension; mole fraction of oxygen; extent of reaction
x_o	mole fraction of oxygen at outlet of reactor
Y	yield of reaction
Y_A	mole fraction of species A
y	weight % of coke on catalyst; mole fraction
Z	length along reactor
z	$l/d_p Z$
Da	Damköhler number
Le	Lewis number, $D_e \rho c_p / K_e$
Nu	Nusselt number, $h d_p / K$ [defined in equation (3.3)]
Nu*	modified Nusselt number, $h d_p / K_e$
Pr	Prandtl number, $c_p \mu / K$
Re	Reynolds number
Sc	Schmidt number, $= \mu / \rho D$ [equation (3.4)]
Sh	Sherwood number, $k_c d_p / D$ [defined in equation (3.3)]
Sh*	modified Sherwood number, $k_c d_p / D_e$

Greek symbols

α	constant defined in equation (3.11); geometric factor for catalyst pellet
β	thermicity factor
β_c	$\Delta H_c C_{og} / (c_p T_g)$
β_H	$D_e \Delta H_2 C_{og} L_e / (K_e T_g)$
γ	Arrhenius parameter, E/RT_0; parameter defined in equation (9.70)
γ_d	Arrhenius parameter for deactivation
γ_1, γ_2	metal support and metal atmosphere interfacial energies in sintering
δ	dimensionless pellet radius, r/R; ratio of cell height to particle diameter, cell model

ε	void fraction
ε_a	micropore void fraction
ε_i	micropore void fraction
ε'	void fraction in bed
η	effectiveness factor
η_c	R_c^*/R_{ref}
η_{H_2}	R_H^*/R_{ref}
θ	dimensionless temperature, T/T_0; fraction of active sites; time
θ_o	initial fractional area of active sites; time interval
θ_p	poisoned fractional area of active sites
θ_∞	ratio of total capacity of catalyst for poison adsorption to feed rate of poison
λ	effective reaction zone thickness in catalyst pellet; deactivation parameter
λ_d	dimensionless radius of sharp interface, λ/R
μ	viscosity
ξ	dimensionless radius of catalyst pellet; dimensionless reactor length
ρ	density; dimensionless reaction radius in unreacted core models
ρ	dimensionless radius of reacting retracting core models
ρ^*	dummy variable; dimensionless radius of reacting interface in retracting core models
ρ_b	bulk density
ρ_F	density of feed
ρ_g	gas or fluid density
ρ_s	solid density
σ	characteristic parameter for fouling, $D_{eA}/\varepsilon R^2 k_{fl} C_{Ao}$; ratio of fluid concentration gradient at the diffusion reaction interface to the linear gradient across the reaction zone—finite reaction zone model
σ_{AB}	collision diameter of two molecules A and B
τ	tortuosity; dimensionless time
τ_p	poisoning time
Φ	deactivation function
Φ_{kj}	rate of collision between surface crystallites
ϕ	Thiele modulus
ϕ_a	$(3Sh^*/C_{Ag})(C_{Ag} - C_{As})$ [equation (9.15)]
ϕ_d	Thiele modulus under deactivating conditions
ϕ_o	$R(k_c C_{Bo}\sigma/6nD_e)^{1/2}$ [equation (9.13)]
ϕ_p	Thiele modulus for poisoning reaction
ψ	spreading pressure in Kelvin equation; fractional coverage of surface by poison, q/q_0; dimensionless oxygen concentration, C_o/C_{og}; activity function; parameter defined in equation (9.69)
Ω	deactivation function; reaction modulus
Ω_D	collision integral

Subscripts

e	exit conditions
F	feed

f	refers to fouling
g	gas phase
i	interface, subscript in cell model
j	subscript in cell model
o	initial or bulk conditions
P	poison

1
Introduction

1.1 General introduction

The use of catalysts in the chemical and process industries is currently widespread. The nature of the catalytic process whereby a given reaction may be accomplished at a lower temperature than that required for the homogeneous reaction is likely to lead to the application of catalytic methods to even more processes as fuel costs continue to rise. Unfortunately most catalysts used in heterogeneous catalytic processes are subject to a decrease in the initial activity over a period of time. The time required for the activity of a catalyst to fall to an undesirable level varies with the severity of the process conditions and with the type of reaction being catalysed. Thus for catalytic cracking of petroleum feedstocks the catalyst activity may fall to an unacceptable level after only a few seconds contact time whereas for the synthesis of ammonia and the reforming of naphtha over a platinum catalyst the active life of the catalyst may be of the order of 1 year or more. This wide variation in the time that a catalyst may be employed efficiently has an important bearing on both the design and the operation of catalytic reactors. If the time during which a catalyst decays is very short, some form of continuous regeneration of the catalyst is required and this leads to the adoption of either a fluidized or a moving bed reactor for the process. On the other hand, if the catalyst life were (say) 1 year, a fixed bed reactor (or reactors) with periodic regeneration might be preferred. In this latter case, if the catalyst is not too expensive it may be discarded after 1 or 2 years operation and the reactor recharged with fresh catalyst.

The complete process of converting a given feedstock into a required product stream in a catalytic reactor is complicated and involves chemical and physical steps occurring both simultaneously and in sequence. Of prime importance is the chemical activity of the catalyst. This must be adequate to fulfil the requirements of the process and, as indicated above, the catalyst activity should have a reasonable life, if possible. However, chemical considerations alone are not sufficient. In fixed bed operation the pressure drop must be kept within specified bounds, so catalysts used in this type of reactor are usually in the form of granules or pellets. Therefore the reactants have necessarily to diffuse through the porous matrix of these granules or pellets to reach the active sites of the catalyst. This imposes a mass transport limitation on the intrinsic chemical rate. External mass transport from the fluid stream to the pellet surface may also be important in some cases. This last effect is usually minimized in commercial operation by the use of high fluid flow rates, but the associated heat transfer restriction frequently cannot be removed by this means and there may be significant temperature differences between the bulk fluid and the external surface of the catalyst pellet. Heat effects may also be associated with the pore structure of the pellet. The complex pore structure present in most catalysts may restrict the transport of reaction heat, thus leading to intraparticle temperature gradients.

The necessity for having the catalyst in pelleted form has a bearing on catalyst deactivation when this is caused by deposition of foreign matter. If such a deactivation process is occurring this can lead under some conditions to pore blocking with consequent loss of active catalyst area, and in extreme cases it may even lead to blocking of the interparticle void spaces and complete blockage of the reactor. The employment of pelleted catalysts may, however, be beneficial in some cases since, if the deactivation of the catalyst occurs first at the outer surface of the pellet, there will be "reserve" activity within the pellet which lessens the overall deactivating effect. Thus any consideration of catalyst deactivation and its applications must include an assessment of the physical processes and their relation to the structure of the pelleted catalyst. These in turn influence the overall reactor behaviour and determine the optimal operating policy to minimize the effects of deactivation and affect the regeneration procedure.

As might be expected, because of its industrial importance, the subject of catalyst deactivation has received a great deal of attention. Nevertheless there is some confusion apparent in the literature between the effects of different forms of deactivation, and frequently results relevant to one form of deactivation are applied quite erroneously to other forms of deactivation. This sometimes has led to false predictions being made. Additionally, in the past few years there has been significant progress in the understanding of the phenomenon of catalyst deactivation, so understanding of the processes involved is now much clearer.

The present monograph has the aim of bringing together both the chemical and chemical engineering aspects of catalyst deactivation. Although the individual chemical and physical processes, such as chemical poisoning and thermal sintering of the catalyst, will be reviewed, these will also be considered in the context of the overall problem of reaction plus diffusion in catalyst pellets under deactivating conditions. These in turn must be related to the problems of reactor behaviour and the best operating policies to be selected to minimize the effects of catalyst deactivation. These problems are discussed later, together with the means whereby catalyst activity is restored by regeneration.

1.2 Classification of catalyst deactivation processes

General reviews of catalyst deactivation have been given by Butt (1972) and by Levenspiel (1972) and they have laid the foundations of a better understanding of catalyst deactivation processes. Basically, three kinds of deactivation may occur: (1) sintering or thermal deactivation of the catalyst; (2) poisoning; (3) fouling.

Sintering is a physical process associated with loss of area of the catalyst which occurs when the catalyst is operated above the normal range of temperature. Such temperature rises may occur throughout the catalyst or may be localized at the individual areas where reaction occurs. Two different kinds of sintering may be distinguished depending on the type of catalyst employed. If the catalyst is a normal high-area support type material, such as SiO_2 or the various forms of Al_2O_3 or a silica–alumina cracking catalyst, operation at high temperatures will cause a loss of specific surface with associated changes in the pore structure, giving a corresponding loss in activity. The second type of catalyst is that where the active ingredient is usually a metal which is supported on a high-area oxide support. Examples are the nickel and platinum catalysts supported on alumina or silica. Here sintering can occur not only by reduction of the support area but by a "coalescence" or loss of dispersion of the metal crystallites. This loss of area of the active constituent of the catalyst causes a sharp drop in activity; furthermore this type of sintering may occur at temperatures below that at which the support material suffers loss of area. In the case of a bifunctional catalyst such as the platinum on alumina reforming catalysts, temperature excursions may reduce the area of both components. In other cases where the support does not have a direct chemical role one of its main functions is to prevent aggregation of the metal crystallites.

Poisoning was once the generic name applied to all forms of catalyst deactivation and unfortunately is still often used in this non-specific sense. In this monograph the term poisoning will be used solely to describe catalyst deactivation due to small amounts of material, specific to a specific catalyst and associated with the adsorption of the poison on the active sites of the catalyst. Poisoning is often associated with contaminants such as sulphur compounds in the feed stream of petroleum fractions; it is then termed impurity poisoning. Though this is the most well documented and best identified of poisoning processes, it is important to recognize that other forms of poisoning may occur. These include poisoning by a product of the desired reaction which may be preferentially adsorbed on the active sites of the catalyst, thus retarding the adsorption of reactant.

Most poisoning processes are effectively irreversible, so the catalyst has to be discarded ultimately, but there is an important class of poisons that are reversible in action. Examples include the poisoning of catalysts by oxygen and nitrogen (Maxted, 1951) and the poisoning of nickel hydrogenation catalysts by water vapour (Gioia, 1971) and by oxygen (Koh and Hughes, 1974). In principle it is always possible to remove impurity poisons from the feed stream by purifying the feedstock or by using a guard catalyst, but in practice the cost of purifying the feedstock to achieve the very low impurity level often required (less than 1 ppm of sulphur compounds for nickel methanation catalysts) may prove prohibitive, so it may often be preferable to tolerate some degree of poisoning. Fortunately, for many fixed bed catalytic reactors the reaction zone occupies only a small fraction of the total bed length. The reaction zone is also the site of the poisoning process, so the particular region of the bed where the reaction is occurring ultimately becomes deactivated. The reaction zone then moves downstream to the next active section of the bed. This process continues with a continuous movement of the reaction zone from the reactor inlet to the exit, with the deactivated catalyst being upstream of the active catalyst throughout. By having sufficient bed length, adequate catalyst life may be obtained. One example is the methanation process where catalyst replacement is necessarily only at intervals of about 1 year.

It should be emphasized that poisoning is not always undesirable; in some cases selective poisoning may be employed to enhance one reaction on a multifunctional catalyst whilst inhibiting a less desirable one. In this particular case, selectivity enhancement is achieved by preferential blocking of certain sites on the catalyst. In other cases selectivity improvement may be obtained by reducing the deleterious effect of the diffusional limitation on the desired process.

Fouling is a process of catalyst deactivation that may be either physical or chemical in nature. In general, much larger amounts of material are responsible for deactivation in fouling processes than in poisoning. The most typical of fouling processes is that of the carbonaceous deposit or "coke" that forms on most catalysts used in the processing of petroleum fractions or other organic chemical feedstocks. Another class of fouling reactions is that of metal sulphide deposition arising from the organometallic constituents of petroleum which react with sulphur-containing molecules and deposit within the pores of the catalyst during hydrotreating operations (Newson, 1975). Liquid fuels derived from coals also give rise to similar deposits during hydrocracking of the resultant liquid (Davies, 1978). These last two examples may be termed impurity fouling, and as such they are not typical of the more general type of fouling associated with coke formation on the catalyst surface. The formation of carbonaceous or coke deposits (containing, in addition to carbon, significant amounts of hydrogen plus traces of oxygen, sulphur, and nitrogen) during the processing of organic based chemical feedstocks is the more usual example of fouling. It is important to recognize that the coke deposit in this case originates from the reactions occurring and is not an impurity. Because of this intrinsic association with the main chemical reactions, fouling by coke cannot be eliminated by purification of the feed or use of a guard catalyst; if reaction occurs, coke deposition must also necessarily occur according to the overall chemistry of the process. However, coke formation can be minimized by appropriate choice of reactor and operating conditions, and in some cases by modification of the catalyst.

In general, coke can originate from the reactant or product by reactions (1) or (2). Reaction (1) is called parallel fouling while (2) is called series fouling or consecutive fouling. Fouling may (and indeed often does for complex systems) also occur by a combination of reactions (1) and (2). The extent of coke formation will depend on the orders of reaction with respect to the formation of the desired product R and the coke, and on the magnitudes of the temperature coefficients for each reaction.

$$A \begin{cases} \rightarrow R \text{ (main reaction)} \\ \rightarrow \text{Coke (side reaction)} \end{cases} \quad (1)$$

$$A \rightarrow R \rightarrow \text{Coke} \quad (2)$$

An important distinction between fouling and poisoning is provided by a study of sulphur poisoning of Fischer–Tropsch catalysts carried out by Anderson and coworkers (1965). They observed that a decrease of catalyst particle size caused less catalyst deactivation, and this was attributed to the increased external surface due to this subdivision of the original particles being able to accommodate the poison without too much activity loss, since the poisoning occurred in a thin active layer near to the external surface of the particles. Similar reasoning would not be expected to hold for deactivation by coking, since coke deposition must necessarily occur at points where the reaction rate is greatest, and this occurs whether the coke precursor is the reactant or the product. In the case of parallel fouling it can be shown (Kam et al., 1975) that the best use of the catalyst is obtained by increasing the particle size, in contrast to the above example for impurity poisoning.

Fouling is associated with relatively large amounts of deposit and, if it is excessive, in addition to covering the active sites of the catalyst it may affect the diffusional properties of the porous catalyst pellet. In the limit, pore blocking may occur for both coke and metal deposition, and if allowed to continue the deposits will block the void spaces between the catalyst pellets, necessitating a complete shut-down of the reactor.

An important operating characteristic of any commercial catalyst is that of catalyst strength. This must be retained throughout the life of the catalyst, otherwise pellet disintegration will occur and the severe pressure drop resulting from the catalyst breakdown may make it necessary for the reactor to be shut down. Catalyst break-up can occur with severe catalyst fouling alone but more often it is associated with a combination of severe catalyst fouling and extreme operating conditions of the catalyst. These may be physical, such as extreme (and frequent) temperature cycling, or due to maloperation of the process which may include extreme feed variation and wrong sequencing of the feed stream. An exhaustive discussion of catalyst strength is outside the bounds of this text but the subject will be referred to when loss of strength is a consequence of catalyst deactivation.

In the following chapters the general development of catalyst deactivation processes will first be reviewed, followed by sections devoted to the specific deactivation mechanisms noted above. The effects on catalytic reactors will then be considered, and finally operating policies and regeneration procedures for reactors operating under conditions where catalyst deactivation is important will be assessed.

References

Anderson, R. B., Karn, F. S. and Shultz, J. F. (1965). *J. Catal.* **4**, 56.
Butt, J. B. (1972). *Adv. Chem. Ser.* **109**, 359.
Davies, G. O. (1978). *Chem. and Ind.* 560.
Gioia, F. (1971). *Ind. Eng. Chem. (Fund.)* **10**, 204.
Kam, E. K. T., Ramachandran, P. A. and Hughes, R. (1975). *J. Catal.* **38**, 283.
Koh, H. P. and Hughes, R. (1974). *A. I. Ch. E. J.* **20**, 395.
Levenspiel, O. (1972). *Chemical Reaction Engineering*, 2nd Edn, Ch. 15. John Wiley, New York.
Maxted, E. B. (1951). *Adv. Catal.* **13**, 129.
Newson, E. J. (1975). *Ind. Eng. Chem. (Proc. Des. Devel.)* **14**, 27.

2
General Aspects of Catalyst Deactivation

2.1 Poisoning of catalysts

2.1.1 Poisoning of metallic catalysts

The "life" of a catalyst may be defined as the period during which the catalyst produces the required product at a yield equal to or greater than that originally specified. For most catalysts the activity declines sharply at first and then reaches a state where the catalyst activity decreases much more slowly with time. The selectivity associated with the activity change may improve or become worse. It is with the second or slow stage that we are interested mainly, although with some catalysts the cycle time is so short that the first rapid deactivation stage becomes of necessity of prime interest.

Earlier work on the poisoning of metallic catalysts was brought together by Maxted (1951) in an excellent review. In this he proposed a theory of catalyst poisoning based on the electron structural properties of the poison in the gas phase and the solid metal catalyst. The concept employed is that essentially the poison is adsorbed on the active metal sites to form a chemisorbed complex. Limitation to chemically bonded systems implies a specificity and also that a low concentration of poison may have a very marked deactivating effect. Metallic catalysts susceptible to poisoning are confined mainly to metals of Group VIII of the periodic table and the closely related metals of Group IB (Cu, Ag, Au). These are listed in Table 2.1; most of them are employed for hydrogenation and reforming reactions, and much of the earlier work on poisoning was performed on this type of catalyst.

Table 2.1 Catalytic metals most susceptible to poisoning.

Fe	Co	Ni	Cu
Ru	Rh	Pd	Ag
Os	Ir	Pt	Au

The principal poisons that are effective in deactivating these metal catalysts belong to the following groups:

(a) Molecules containing elements of the periodic table Groups VB and VIB, i.e. N, P, As, Sb, and O, S, Se, Te.
(b) Compounds of a large number of catalytically toxic metals.
(c) Molecules containing multiple bonds, such as CO, cyanogen compounds, and strongly adsorbed organic molecules.

The toxicity of the compounds of Groups VB and VIB was attributed by Maxted to the presence of unshared electron pairs which facilitated the chemisorption process. Thus compounds such as H_2S, PH_3, and organic sulphides function as poisons whereas compounds with no lone electron pair (e.g. sulphonic acids) are non-toxic.

An interesting aspect of Maxted's work that is often overlooked is his study of detoxification of catalysts. In Maxted's work this essentially consisted of washing the catalyst with a detoxifying agent. Although the use of liquid reagents would not be desirable in commercial practice, application of some of these treatments in the vapour phase might prove effective and would present an alternative to discarding the catalyst and reprocessing it to manufacture a new catalyst.

Catalyst poisons containing toxic metals were extensively investigated by Paal and coworkers (1913, 1918) mainly for their effect on platinum and palladium catalysts. It was found that many heavy metals, including mercury, lead, bismuth, tin, and zinc, cadmium, and copper, could reduce the activity of the catalyst. In a more systematic investigation of this effect, Maxted and Marsden (1946) concluded that the toxicity of the heavy metals was associated with all 5d electron orbitals being occupied by electron pairs or containing at least one electron in the d orbital. No toxicity was observed if no d orbitals were present or if the d orbitals were unoccupied. The d electrons therefore seem to be responsible for the intermetallic bond between the metal and the catalyst and thus cause toxicity.

Poisons that owe their toxicity to the presence of multiple bonds present a particularly interesting case since these unsaturated molecules constitute the principal class of substances usually hydrogenated using transition metal catalysts. A very wide spectrum of poisoning efficiency is observed, however, for different unsaturated compounds. Thus, if two unsaturated substances are present during the same hydrogenation process, the retarding effect of one component on the hydrogenation of the other may vary from mere competition for the active sites of the catalyst to almost complete suppression of one reaction by the presence of even small amounts of the other. This effect may be used to advantage in systems where one component is to be selectively hydrogenated.

2 General Aspects of Catalyst Deactivation

The most well known example of an unsaturated compound that causes poisoning is carbon monoxide. It should be emphasized that the toxicity of carbon monoxide is lost if the degree of unsaturation is reduced. Thus carbon monoxide when oxidized to carbon dioxide loses its poisoning effect.

2.1.2 Poisoning of non-metallic catalysts

Most reported work on the poisoning of non-metallic catalysts refers to cracking catalysts. These are usually acidic oxides which perform the isomerization, cracking, and double-bond migration reactions constituting the catalytic process. It is now well established that basic nitrogen compounds are poisons for these catalysts. Most cracking catalysts are based on alumina, silica, or silica–alumina. The latter is the most common and may be used in the so-called amorphous form or as synthetic zeolites embedded in a silica–alumina matrix. Early work by Mills *et al.* (1950) on silica–alumina catalysts used for cumene cracking established that the poisoning effectiveness of basic organic nitrogen compounds could be ranked in the order: quinaldine > quinoline > pyrrole > piperidine > decylamine > aniline. The surprising feature of this order of poisoning effectiveness is that it does not correspond to the order of basicity which might be expected since a stronger base would be expected to neutralize more of the acid sites. In fact piperidine is the strongest base, yet as the order above shows it is by no means the most effective poison. This apparent anomaly is resolved when account is taken of the cracking occurring within the nitrogen compounds. When this is done it is found that piperidine is 54% cracked to products (Mills *et al.*, 1950), and consequently its poisoning effectiveness must be decreased. In addition, Mills *et al.* obtained a correlation between catalyst activity for cracking catalysts and equilibrium quinoline adsorption. This is illustrated in Fig. 2.1.

The mechanism of poisoning by these organic bases is postulated to occur by chemisorption of the poison on aluminium or silicon ions in the catalyst which are not fully coordinated. The adsorbed complexes form Lewis acids on the surface.

Mills *et al.* (1950) also established that, for the deactivation of cracking catalysts for cumene cracking, all the nitrogen compounds were selective

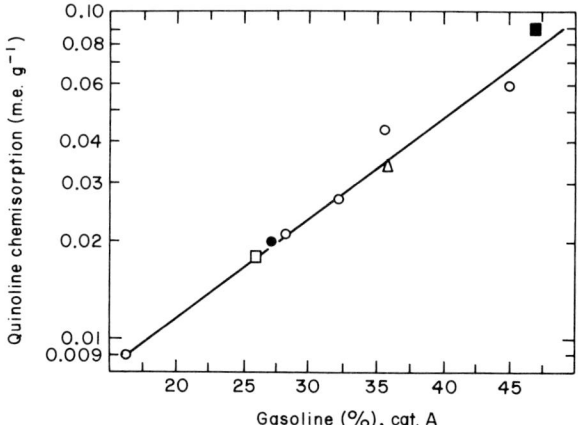

Fig. 2.1 Quinoline chemisorption at 315°C as a function of activity for cracking light East Texas gas oil: ○ SiO$_2$–Al$_2$O$_3$ (Houdry type S); ● SiO$_2$–1% Al$_2$O$_3$; □ clay catalyst (Filtrol); ■ SiO$_2$–MgO; △ SiO$_2$–ZnO. (Mills *et al.*, 1950)

poisons, with an approximate exponential dependence of activity on the poison concentration. Thus, for example, a 4% coverage of a silica–13% alumina catalyst by quinoline caused a 7-fold decrease in the catalyst activity. It may be concluded from this result that only a small fraction of the surface participates in the cracking activity of the catalyst.

This suggestion of a distribution of acid strengths on cracking catalysts was investigated in detail by Pines and Haag (1960); model reactions requiring differing degrees of acid strength for successful reaction were carried out on a series of aluminas. The three types of alumina studied were:

(a) Pure alumina; prepared from aluminium isopropoxide or from Al(OH)$_3$ and Al(NO$_3$)$_3$.
(b) Impregnated; pure catalyst as in (a) impregnated with NaCl or NaOH.
(c) Aluminate; alkali containing alumina precipitated from KOH solution.

The incorporation of alkali into catalysts (b) and (c) decreased the activity compared with the pure catalyst (a), and thus enabled a varied acidity spectrum to be studied.

The results of activity studies were in good agreement with predictions based on the relative acidity of the various catalysts. The pure alumina (a), prepared from isopropoxide solutions, gave a very high activity for –CH– isomerizations, while the alkali doped catalysts showed very little activity for the same reactions. Preferential poisoning of strong acid sites was shown by various organic bases, and Pines and Haag (1960) calculated that pure

2 General Aspects of Catalyst Deactivation

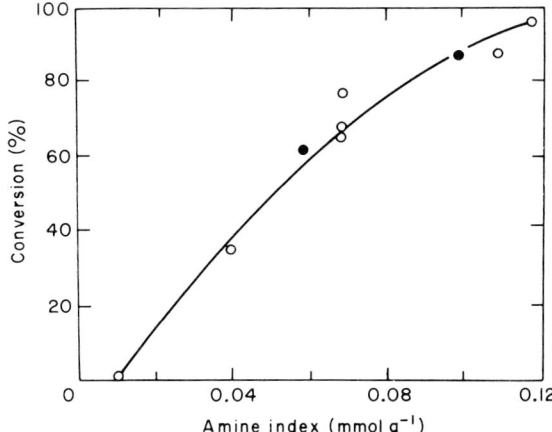

Fig. 2.2 Conversion of 3,3-dimethyl-1-butene (350°, HLSV 2.0) *vs.* acidity; alumina from isopropoxide: ○ NaOH impregnated; ● NaCl impregnated. (Pines and Haag, 1960)

alumina contained about 10^{18} acidic sites m^{-2}, of which probably 10% are effective in isomerization reactions.

In the same way as Mills *et al.* (1950) were able to correlate activity with equilibrium quinoline adsorption, Pines and Haag (1960) demonstrated that an additional relationship could be obtained between amine index and catalytic activity. The amine index is another method of estimating acidity and, as Fig. 2.2 shows, it enables a comparison to be made of the relative activities of various aluminas.

The important practical result from this work on the doping of catalysts with alkali is that it has led to improved catalyst formulation for reactions where by-product formation of coke is appreciable. As will be seen later when coking processes are considered, the addition of alkali to suppress the acidity of a support, and hence diminish the extent of cracking reactions, has a beneficial effect in reducing the extent of coke deposition on a catalyst. Thus addition of small amounts (usually 1% or less) of alkali to nickel on alumina steam reforming catalysts reduces coke formation and extends the useful life of the catalyst (Bridger, 1970).

Other investigations on the poisoning of cracking catalysts by organic bases have been made by Parera (1968). For the poisoning of the methanol dehydration reaction over a commercial silica–alumina catalyst with various organic bases containing nitrogen, a plot of activity against poison concentration was linear for the first 60–80% of the total activity decrease. After this the activity fell off much less steeply with further increase in poison concentration. Parera showed that the linear region in these experiments was due to the poisoning of acid sites of uniform activity. The part of the activity curve corresponding to high concentrations of poisons could be indicative of poison adsorbed on adjacent sites at high surface coverage. A further result, quoted by Parera, is that the amount of organic base required for complete deactivation of the catalyst did not coincide with the classical value of the acidity, determined at room temperatures by the butylamine index. Even with the weakest poison employed (pyridine), only half of the amount predicted from the butylamine index gave complete deactivation.

With increasing use of zeolite catalysts for hydrocarbon processing, and especially for catalytic cracking, attempts have been made to measure the acidity of zeolitic catalysts by quinoline titration. The difficulties involved in such measurements have been demonstrated by Goldstein and Morgan (1970) who investigated the quinoline titration of Y type zeolites. The reaction investigated was cumene cracking in the temperature range 250–450°C. It was shown that the maximum titre of quinoline, which might be expected to correspond to the concentration of active sites on the catalyst surface, did not correlate in this manner. Instead the minimum titre corresponded with the density of supercages in the structure. These supercages are a feature of the structure of zeolites and impart the well known sieving or shape selective properties to the zeolite. It seems that a single molecule of quinoline within the supercage is sufficient to block cumene from the active sites. The danger of misinterpretation of quinoline titre results with this type of catalyst structure is obvious.

2.1.3 Poisoning of bifunctional catalysts

The poisoning of metallic catalysts was discussed in Section 2.1.1, while similar behaviour for non-metallic catalysts (mainly oxides) was reviewed in Section 2.1.2. A large number of catalysts, however, are composite in nature in that they consist of a metal dispersed as very small crystallites on a large area support. The support is usually in the form of a refractory oxide, and in addition to providing the high area on which the metal is dispersed it also has the action of a "spacer" or stabilizer, preventing the highly active metal crystallites from coalescing (sintering). Although in some cases the support is inert chemically to the catalytic process operation, in many cases it will also have a catalytic function. Examples of this type of bifunctional catalyst are abundant in the petroleum processing industry, a classic example being platinum reforming catalysts used to produce gasoline from naphtha (Ciapetta et al., 1958; Fowle et al., 1951). These reforming catalysts contain 0.3–1.0% of platinum on an acidic alumina (γ or η) support. The dehydrogenation reaction occurs on the platinum sites while isomerization occurs on the alumina. For this type of catalyst, if one component is poisoned more effectively than the other, the selectivity of the catalyst will be affected as well as the activity. An experimental determination of the poisoning of bifunctional catalysts was made by Webb and McNab (1972) on a rhodium–silica catalyst. The addition of mercury, which from the work of Maxted is well known as a poison for the transition metals, caused a considerable diminution in the hydrogenation activity of the rhodium whilst scarcely affecting the isomerization rate on the alumina sites. These results support the view that hydrogenation and isomerization reactions proceed independently of each other on the metallic and acidic portions of the catalyst surface respectively.

2.2 Coke formation on catalysts

Coke formation on catalysts may take several forms and be caused by different mechanisms. If the temperature is high enough pyrolysis reactions may occur in the gas phase and deposition of "gas phase carbon" may occur on the catalyst. At lower temperatures carbon or coke is usually formed directly on the catalyst surface by means of the catalytic processes. In its simplest form this may occur by a parallel or series reaction to the main reaction as mentioned in Chapter 1. Recent work using advanced techniques, including surface examination by electron microscopy (Trimm and co-workers, 1977) and controlled atmosphere electron microscopy (Baker and Chludzinski, 1980), has established that, with metals and supported metal catalysts, coke deposition may occur in the form of filaments. This raises some interesting questions concerning the mechanism of carbon formation under these conditions, and also how this type of growth may affect the transport of fluid within the catalyst. However, for oxide catalysts, including cracking catalysts, Haldeman and Botty (1959), using microscopic techniques, concluded that the coke deposit is in the form of thin filmy aggregates of particle size less than 10 nm. An important result was that for a 1–2% deposition of coke on a cracking catalyst (which was approximately the level at which regeneration became necessary) complete coverage of the surface by coke is not attained. This confirms the non-uniformity of the active sites on the surface and also the aggregation of the coke into particles. The value of 10 nm for the aggregate size agrees broadly with estimates made by Massoth (1967) for pore blocking by coke aggregates where a size of about 4 nm is common.

The species acting as precursor for the coke has been the subject of much investigation and speculation. Earlier work by Blue and Engle (1951) suggested that olefins were largely responsible for coke deposition at low conversion but aromatics contributed significantly to deposition at higher conversions. However, Eberly *et al.* (1966) have found from infrared studies of the surface that aromatic skeletal vibrations are present. From this they suggested that coke formation involves the initial adsorption of hydrocarbons followed by chemical reaction on the surface. The chemical reactions occurring include condensation followed by hydrogen elimination. It is probably true to say that this problem of the immediate precursor of coke is still unresolved. On the whole, the majority of opinion is in favour of aromatic species, but there is some strong evidence that olefins can be important.

Quantitative investigations started with correlations, usually empirical, of the amount of coke deposition as a function of time. Some earlier reforming studies of naphtha to gasoline over a molybdena–alumina catalyst by Wilson and Den Herder (1958) established that coke formation increased with Research Octane Number, which for the system used means that coking increased with temperature. Perhaps the most well known work is that of Voorhies (1945) who studied the coking of fresh and regenerated natural and synthetic cracking catalysts. Some typical results are illustrated in Fig. 2.3. The correlation obtained was:

$$C_c = At^n \tag{2.1}$$

where C_c is the concentration of coke on the catalyst, t is the process time, and A and n are constants. A wide variety of catalysts and operating conditions were employed to obtain this correlation, and Voorhies's results

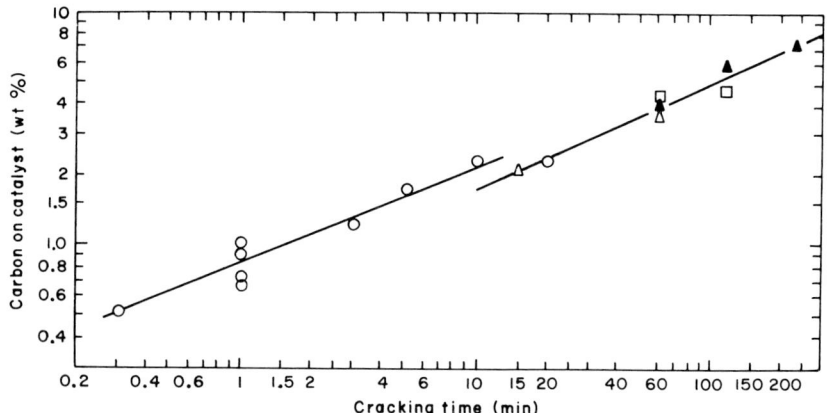

Fig. 2.3 Carbon formation *vs.* cracking time in fixed-bed catalytic cracking. West Texas gas oil (natural catalyst): ○ 1.2 v/v h^{-1}. East Texas gas oil (synthetic catalyst): ▲ 0.3, □ 0.6, △ 1.2 v/v h^{-1}. (Voorhies, 1945)

2 General Aspects of Catalyst Deactivation

indicated that, for fixed bed operation, the coking rate was independent of space velocity and also had a very low dependence on temperature. The value of the exponent n varied somewhat but was in the range 0.5–1. This value of n less than 1 and the small temperature dependence of the coking rate has led to the postulate that catalyst fouling by coke deposition is diffusion controlled.

The lack of dependence of coke deposition on space velocity and the postulate of diffusion control have caused a great amount of criticism, and attempts have been made to resolve these problems experimentally. Eberly and coworkers (1966) made an extensive study of coke formation on silica–alumina cracking catalysts, part of which has already been referred to in connection with the mechanism of coke formation. A major portion of this same study was concerned with the effect of operating variables. It was found that coke formation in fixed beds was a complex function of length of the cracking period and the feed rate. The coke deposition rate was found to have some dependence on space velocity as Fig. 2.4 shows. A maximum in the coke deposit with space velocity was obtained which indicated that the carbon laydown was not uniform through the bed. These maxima were interpreted as being caused by the complex nature of the coke formation mechanism involving adsorption followed by surface reaction. A more detailed examination of the coke deposition *vs.* space velocity curves obtained by Eberly *et al.* (1966) in Fig. 2.4 does reveal that after initial

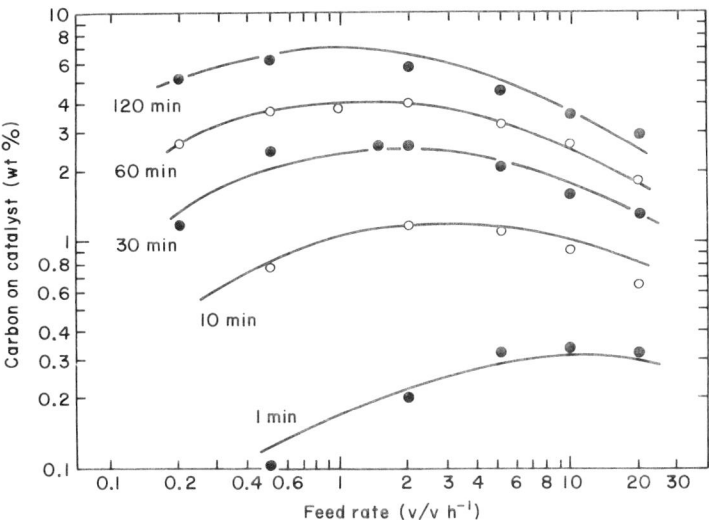

Fig. 2.4 Carbon formation on $13\% \text{Al}_2\text{O}_3$–$87\% \text{SiO}_2$ from cracking of n-hexadecane at 500°C. (Eberly *et al.*, 1966)

operation the curves are almost flat, so bearing in mind the more restricted range of space velocity used by Voorhies (1945) it is easy to see how the lack of dependence of coke deposit on space velocity was concluded. Blue and Engle (1951) also found that the coking rate was almost independent of space velocity. It may be assumed therefore that the Voorhies correlation may be used for a large number of reactions where deactivation by coking occurs, and this is borne out in practice.

The question of diffusion control in the coking of a catalyst is more open to criticism. The whole problem of diffusional processes in porous catalyst pellets will be discussed in the following and later chapters where it will be seen that the subject is very complex and depends on the operating conditions to a great extent. However, attempts have been made to check this assumption by several workers. Eberly *et al.* (1966), whose work on mechanism and space velocity effects has already been referred to above, observed that in their experiments there was no dependence of the coke deposit on particle size and it could therefore be assumed that the coke was uniformly distributed in the catalyst pellets. Rudershausen and Watson (1955) found that, in the aromatization of cyclohexane over a molybdena–alumina hydroforming catalyst, the temperature coefficient of the coking deposition was too high for diffusional control, suggesting that the process is chemically controlled. It would seem therefore that diffusion control may only be present in certain circumstances. This is confirmed by the wide variation in n values obtained. Values of n taken from Eberly *et al.* (1966) and from Ozawa and Bischoff (1968) are given in Tables 2.2 and 2.3. The Ozawa and Bischoff data refer to the second stage ($t > 10$ min).

Table 2.2 Constants in $C_c = At^n$; fixed bed of SiO_2–13% Al_2O_3 at 500°C. (Eberly *et al.*, 1966)

Feed	v/v h^{-1}	A	n
n-Hexadecene	0.2	0.049	0.97
	0.5	0.11	0.86
	1	0.16	0.78
	2	0.22	0.70
	5	0.29	0.60
	10	0.31	0.52
	20	0.29	0.44
East Texas light gas oil	0.5	0.25	0.82
	2	0.52	0.70
	5	1.05	0.72
	10	0.90	0.41

Table 2.3 Exponent in $C_c = At^n$. (Ozawa and Bischoff, 1968)

Reaction temp. (°C)	C$_2$H$_4$ flow rate (ml min^{-1})	n ($t > 10$ min)
499	75	0.885
498	50	0.913
500	30	0.881
450	75	0.742
449	50	0.820
450	30	0.841
400	75	0.625
399	50	0.736
399	30	0.605
350	50	0.554

The usefulness of the Voorhies correlation (2.1) is well substantiated. Many workers have found that coking results can be well correlated using this type of expression. In some cases a single straight line is not obtained when the weight of coke is plotted against time on logarithmic scales, but instead two straight lines are obtained for the initial and later stages of the deposition reaction. Such a situation was obtained by Ozawa and Bischoff (1968) in the cracking of ethylene on a silica–10% alumina cracking catalyst at 500°C. Similar curves have been obtained for the cracking of xylene over a silica–alumina bead catalyst by Hughes and Zadeh (1976). Rudershausen and Watson (1955) obtained the relation:

$$\frac{dC_c}{dt} = \frac{K}{C_c} \qquad (2.2)$$

for the coking rate when investigating the aromatization of cyclohexane. This equation can be integrated easily to give the Voorhies correlation (2.1) with n equal to 0.5, thus demonstrating again the generality of the Voorhies relation. However, in contrast to the work of Voorhies, a high temperature dependence of the coking reaction was observed, and as pointed out above this rules out any diffusional mechanism. More important perhaps, this indicates that K may be related to the actual coking kinetics. If a polymerization process is assumed to be the coking mechanism, as many investigators have proposed, the reaction could propagate by the addition of an adsorbed molecule of reactant to an adjacent coke molecule on the surface to produce a larger coke complex:

$$A_n + A \rightarrow A_{n+1}$$

where A_n is the adsorbed coke molecule, A the adsorbed molecule, and A_{n+1}

the coke product. If this system is then analysed using the usual procedure for a Langmuir–Hinshelwood process controlled by surface reaction between adjacent sites, the rate of coking becomes:

$$r_c = \frac{K'C_{A_n}p_A}{(1 + K_A p_A + K_R p_R + C_{A_n})^2} \qquad (2.3)$$

in which C_{A_n} is the concentration of coke on the surface, p_A and p_R are the partial pressures of coke precursor and reactants respectively, and K_A, K_R, and K' are the usual Langmuir–Hinshelwood constants.

If strong adsorption of coke occurs, the term C_{A_n} is predominant in the denominator of (2.3) and this reduces to:

$$r_c = \frac{K'C_{A_n}p_A}{C_{A_n}^2} = \frac{K'p_A}{C_{A_n}} \qquad (2.4)$$

Under the conditions of the experiments of Rudershausen and Watson, p_A was effectively constant, so equations (2.4) and (2.2) become identical.

Other work on coking has been reported by Appleby and coworkers (1962) who attempted to elucidate the precursors of the coke by studying the cracking of aromatic naphthenic, heterocyclic, and paraffinic hydrocarbons over commercial silica–alumina or silica–zirconia–alumina cracking catalysts. The evidence from this work strongly suggests that aromatic compounds are the coke precursors, those with condensed ring structures such as anthracene being particularly reactive. An important conclusion from this work was that there was an approximate correlation between the extent of coke formation and hydrocarbon basicity. This shows the importance of adsorption strength and provides support for the effectiveness of the catalyst acidity in controlling coke formation as well as the overall reaction.

Since coke deposition is often appreciable (up to 10–20% in some instances) it is conceivable that this may cause pore blocking and thus reduce the accessibility of the interior of the catalyst pellets to the reactant gases. Appleby et al. (1962) observed reductions in catalyst surface area of 22 and 33% with 2.8 and 10.4 wt% coke obtained by the cracking of n-butene and phenanthrene. Ozawa and Bischoff (1968) found no such reduction but the extent of coking was now only about 1%. For non-zeolitic type cracking catalysts this aspect of catalyst fouling has shown little agreement among the various investigators in the past. For example, Ramser and Hill (1958) observed a 27% decrease in surface area for 2.2 wt% of coke while Haldeman and Botty (1959) found almost no effect on either surface area or pore size distribution at coke levels of several wt%. This question has been resolved to some extent by the work of Levinter et al. (1967) who coked a number of different silica–alumina catalysts with styrene–benzene mixtures. A limiting degree of coke formation was observed at which surface area reduction became pronounced. This limiting amount of coke was in general much less than the total amount of coke required to fill the pore space of the catalyst and suggests that pore blocking is preventing utilization of all the internal catalyst area.

The discrepancies between the results of various workers are probably due to the various conditions adopted in the individual experiments and to the different catalysts tested. In terms of pore blocking the quoting of an average wt% of coke on the catalyst is meaningless; it is the distribution of coke deposits that is important. A small amount of coke distributed at the entrance to the pores will have a far greater effect than a more uniform distribution of coke throughout the whole catalyst pellet. Clearly, also the structure of the catalyst pores will be of importance. If the surface pores are of the "ink bottle" type, which either may be an intrinsic feature of the porous catalyst matrix or may be caused by the pelleting operation forming a more closed structure, smaller amounts of coke deposited in the surface layers of the catalyst would be extremely efficient in producing pore blockage. The distribution of coke deposits is also a function of the reaction process occurring and the operating conditions; as will be shown later, a preferential surface deposition of coke can occur under many practical conditions.

For zeolite catalysts, as one might expect, the fine pore structures of the X and Y type molecular sieves or of mordenites are susceptible to pore blockage when coking occurs, and there are no discrepancies such as those observed with the silica–alumina cracking catalyst. Results of Butt *et al.* (1975) on the coking of H-mordenite during cumene cracking show an almost 50% decrease in effective diffusivity for a 4% coke deposit. This last result demonstrates the danger of using effective diffusivity values for fresh catalysts in the estimation of effectiveness factors and other criteria for catalyst performance. If such a procedure is followed, very misleading results will be obtained.

2.3 Metal deposition on catalysts

The contribution of metal deposition to catalyst fouling was mentioned briefly in Chapter 1. Organic metallic compounds exist in many crude oils and in the derived liquids from coal liquefaction. For oils the major metallic impurities are vanadium and nickel while for coal nickel and iron are predominant. In the case of crude oil the metals tend to accumulate in the asphaltene fractions. A comparison of the metal compositions of two crudes is given in Table 2.4.

When these petroleum residua are hydrotreated to remove sulphur, the

Table 2.4 Metals concentration in 560°C + residua. (Oxenreiter et al., 1972)

		Oils	Resins	Asphaltenes
Khafi crude	Composition (wt%)	26	58	16
	Vanadium (ppm)	0	139	736
	Nickel (ppm)	0	36	250
Gach Saran crude	Composition (wt%)	28	62	10
	Vanadium (ppm)	0	282	1455
	Nickel (ppm)	0	100	530

organometallic constituents react out of the oil and combine with hydrogen sulphide to produce solid deposits of metal sulphides (Moritz et al., 1970; Scott et al., 1970). Similar effects occur in the upgrading of the various liquefied coal extracts by hydrotreating. The deposition of metal sulphides may occur either within the pores of the catalyst pellet (intraparticle pore blocking) or between the pellets constituting the bed (interparticle bed plugging effect). Although not directly linked as in catalyst coking, in residuum hydrodesulphurization the desired desulphurization reaction is accompanied always by the parallel demetallation reaction. However, the relative extent of the two reactions is a function of the mean pore diameter of the catalyst, as was demonstrated in some interesting results obtained by Inoguchi (1976). These are shown in Fig. 2.5. For this particular case where

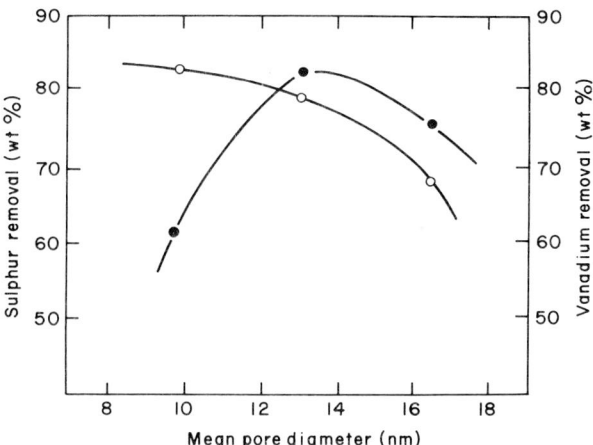

Fig. 2.5 Relation between mean pore diameter and removal of sulphur (○) and of vanadium (●). Feed Ir. Hy.; 410°C; 100 kg cm^{-2}; 1.0 h^{-1}; 1000 H$_2$ 1/1 Oil. (Inoguchi, 1976)

2 General Aspects of Catalyst Deactivation

a heavy residue was treated, desulphurization was shown to be greatest when a relatively small pore size catalyst was employed, whereas demetallation was maximized at a pore size of 12–14 nm. The higher demetallation rate obtained with larger pores has been confirmed by Hastings et al. (1975).

Results from X-ray microanalysis and other techniques have confirmed that, even when the dispersion of cobalt and molybdenum on the support is excellent, large amounts of vanadium, nickel, and sulphur are detected on the outside of the catalyst (Tohdoh et al., 1971). This confirms that the vanadium and nickel are probably deposited as sulphides. Deposition of the metals was found to be influenced by the hydrogen partial pressure. If the hydrogen partial pressure in the hydrodesulphurizer is increased, the metal deposition reaction is increased compared with the desulphurization reaction. Coke deposition invariably accompanies metal deposition, and the procedure often advocated for minimizing coking in hydrotreating units is an increase of hydrogen partial pressure. However, as the results above show (Montagna et al., 1975), this will increase the rate of metal deposition.

The associated problem of bed plugging due to metals deposition in hydrotreating units has attracted attention in recent years. The interparticle effect is usually manifested by increased pressure drop in the bed. Deposition of coke also occurs in the interparticle void space, and the axial distribution of coke and metal deposits in a trickle bed hydrodesulphurizer has been determined. The results obtained are illustrated in Fig. 2.6 taken from the

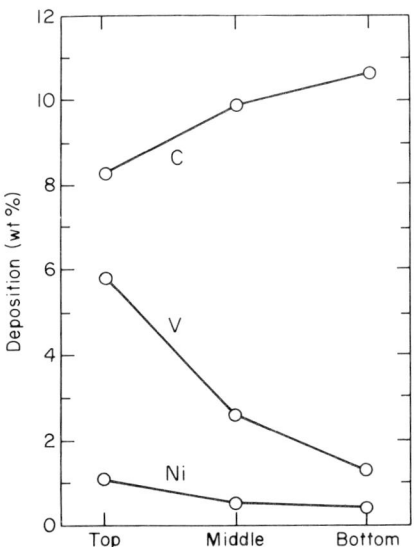

Fig. 2.6 Coke, vanadium, and nickel distribution in the catalyst bed, after 1000 h running. (Kwan and Sato, 1970)

data of Kwan and Sato (1970). Coke deposition in this case is greater in the lower parts of the catalyst bed, i.e. the exit region, while metals deposition occurs preferentially in the entrance region and declines monotonically with distance along the reactor length. Methods for modelling the bed plugging behaviour of metal deposits have been given by Newson (1970, 1975) and he has also suggested methods whereby bed plugging can be minimized. These methods include grading of the beds in terms of both activity and porosity. Many investigators have attempted to forecast the life of a hydrocracking catalyst using methods based on the extent of metal deposition. Frayer *et al.* (1975) and Brunn *et al.* (1976) developed theoretical estimates based on the maximum possible amount of metals deposition and on catalyst porosity and pore size distribution. However, it was observed that the desulphurization activity of the catalyst deteriorated long before reaching these theoretical limits. Ozaki *et al.* (1975) estimated the life of the catalyst based on the kinetics of desulphurization and on an analysis of the reactor temperature gradient, and showed that catalyst activity was lost when 30 g of vanadium were deposited on 100 g of catalyst. The whole area of deactivation of hydrodesulphurization catalysts has been thoroughly reviewed by Ohtsuka (1977).

2.4 Sintering of catalysts

As previously stated, the sintering process may result in an overall loss of area of the support material or oxide base, or may cause a loss of dispersion of the metal crystallites in a supported metal catalyst. Sintering may also be of importance in the reduction of the metal constituent of the catalyst. For example, in the reduction of nickel supported catalysts for steam reforming or methanation, the nickel oxide produced by calcination from the impregnated or coprecipitated nickel salt has to be reduced before the catalyst attains its active state. This process, if not carefully controlled, can lead to loss of area of the nickel crystallites, and this effect is enhanced by the usually exothermic nature of the reduction. Luss (1970) has shown theoretically that very high instantaneous increases in temperature may be achieved on individual metal crystallites under conditions such as these.

Sintering is a complex phenomenon and despite much careful investigation it is still not possible to predict the rate at which the various structural characteristics (e.g. porosity and surface area) change for given operating conditions.

Activation energies, determined experimentally, are high for sintering, so the rate of sintering increases rapidly with increase in temperature. Because of this, for any particular solid there is a temperature, the Tammam temperature, below which sintering is negligibly slow but above which sintering occurs with increasing rapidity as the temperature is increased. The Tammam temperature is usually 0.4–0.5 of the melting point on the absolute temperature scale.

Several mechanisms have been proposed for sintering, some of which are listed below:

(i) Evaporation–condensation. Because the vapour pressure is greater over a convex surface than over a plane or concave surface there will be a tendency for evaporation to occur, from the particles constituting a compacted catalyst pellet to the concave intersections between the particles.
(ii) Volume diffusion. Here diffusion of atoms occurs through the particles to adjacent particles.
(iii) Surface diffusion. This mechanism involves the migration of atoms across the particle surface.
(iv) Grain boundary diffusion.

Whatever the actual mechanism, there is general agreement that for a high area oxide support the sintering process involves three stages. First, there is the growth of the small particle–particle contact areas to form "necks". This is followed by an intermediate stage involving intersection of these "necks" to form closed pores. Finally, if the process is allowed to continue long enough, these closed pores may be eliminated in turn. A pictorial representation of these stages in given in Fig. 2.7.

Inevitably much of the basic mechanistic investigation on sintering has been concerned with powder metallurgy, on which there are a few reviews (Kuczynski et al., 1967; Ristic, 1969) and a number of journal articles (Johnson, 1969; Bache, 1955; Coble, 1967). An example of catalyst sintering is provided by McIver et al. (1963) for aluminas; the results obtained are shown in Fig. 2.8. Loss of both catalyst area and water begins at about 500°C for these materials and becomes serious as the temperature increases.

Much more effort has been devoted to the problem of agglomeration or decreased dispersion of the metal crystallites on a supported metal catalyst. The function of the support (apart from any catalytic activity it may possess, as in a bifunctional catalyst) is to act as a spacer to prevent, or at least minimize, the growth and/or movement of the metal crystallites. However, coalescence will inevitably occur as the temperature is increased; if the support itself sinters, the spacer action will be overcome. Some interesting quantitative results for sintering of both metal crystallite and support have been given by Williams et al. (1972) for a nickel on alumina catalyst. The results obtained are illustrated in Fig. 2.9. Both specific nickel area and total catalyst area drop rapidly in the early stages of the sintering process, and the

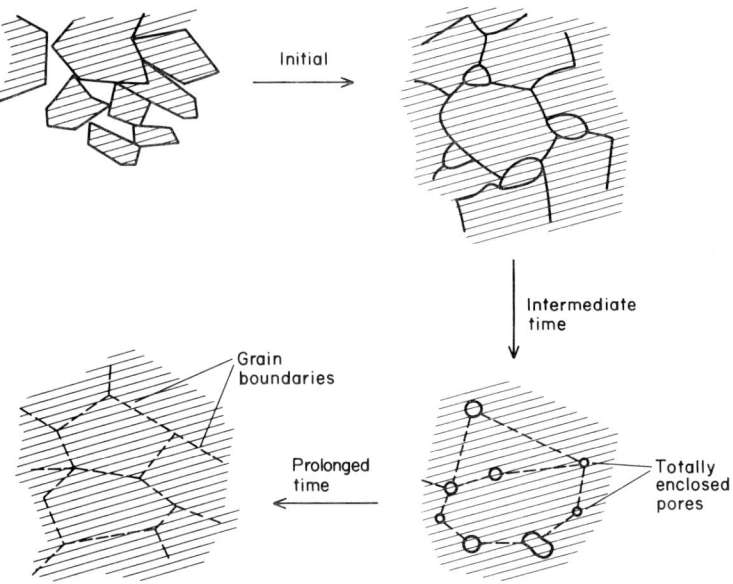

Fig. 2.7 Development of sintering.

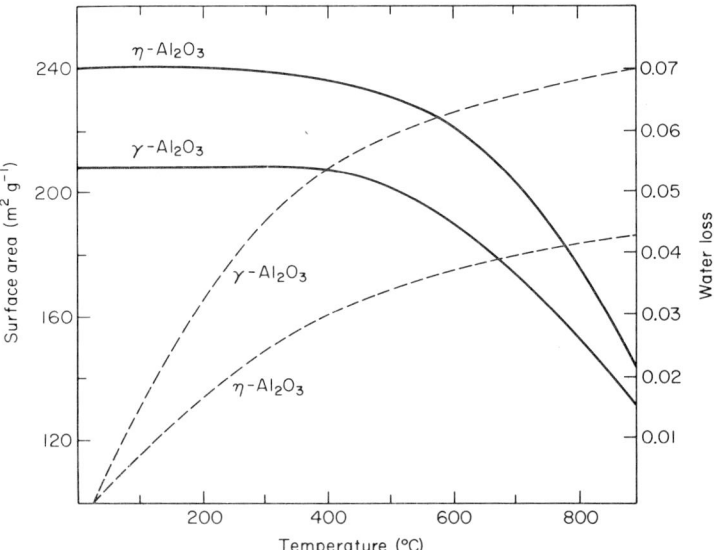

Fig. 2.8 Loss of catalyst area (———) and of water (– – – –) during sintering of two forms of alumina. (McIver *et al.*, 1963)

2 General Aspects of Catalyst Deactivation

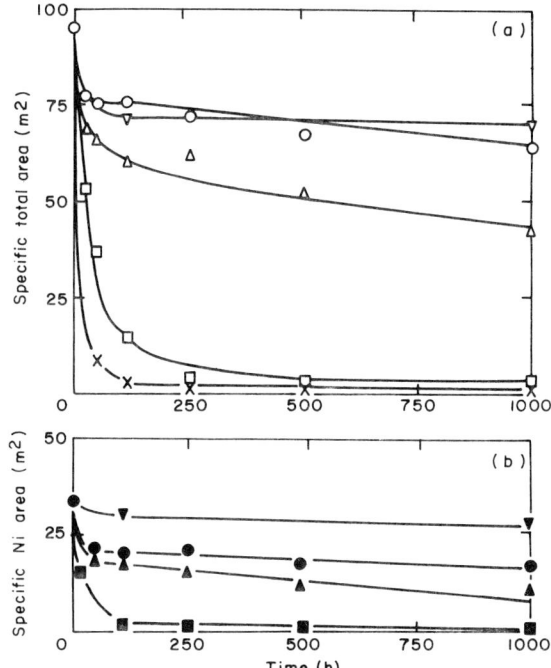

Fig. 2.9 (a) Decrease in specific total area with time in the temperature range 400–800°C: ○ total area at 400°C; △ total area at 500°C; □ total area at 600°C; × total area at 800°C; ▽ total area at 800°C (in hydrogen only).
(b) Decrease in specific nickel area with time in the temperature range 400–800°C: ● nickel area at 400°C; ▲ nickel area at 500°C; ■ nickel area at 600°C; nickel area at 800°C (not measurable); ▼ nickel area at 800°C (in hydrogen only). (Williams et al., 1972)

severity of this initial decrease in area increases as the temperature increases. Following this initial drop, there is only a slow further decrease in area with increasing sintering time.

Most results quoted in the literature for the sintering of the metallic crystallites are for platinum catalysts. A general review has been given by Flynn and Wanke (1975). An interesting feature of platinum on alumina catalysts is that the sintering behaviour appears to be a function of the atmosphere employed. Hermann et al. (1961) were among the first to report details of the change in platinum surface area for Pt/Al_2O_3 catalysts but Hughes and coworkers (1962) in a more quantitative study proposed that the CO adsorption uptake, U, which was used for determining the metal area could be represented by:

$$U = at^b \tag{2.5}$$

By making certain assumptions, this equation could be re-expressed in terms of the dispersion, D, and insertion of appropriate values gave:

$$D = 0.73 t^{-0.13} \tag{2.6}$$

Similarly, Gruher (1962) obtained low values of -0.10 and -0.073 for the exponent in (2.6) for 0.6% platinum and 0.7% platinum on alumina catalysts respectively. These low values of the exponent on the time variable reflect the slow nature of the sintering process.

The empirical data on catalyst sintering have been correlated usually by a power law function of the form:

$$-\frac{dD}{dt} = kD^n \tag{2.7}$$

where k obeys an Arrhenius relationship. Comparison of equation (2.7) with (2.6) and (2.5) shows that a wide variation in power law exponent may be obtained. At first sight the most obvious way of representing a sintering process between metal crystallites on the same support surface is to assume that the process is second-order in metal area. This would give a relation of the form:

$$\frac{1}{A_s} = \frac{1}{A_s^o} + kt \tag{2.8}$$

in which A_s and A_s^o are the current and initial surface areas respectively and k is a sintering constant. The data of Maat and Moscou (1965) on the sintering of a 0.6% platinum on alumina catalyst gave a good straight line when plotted according to equation (2.8).

The possible effects of sintering on selectivity as well as activity are shown by the work of Presland et al. (1972) on the sintering of silver films employed to catalyse the oxidation of ethylene to ethylene oxide. It was observed in this study that an activity decrease due to sintering to 1/8 of the original activity was not accompanied by a corresponding 8-fold decrease in surface area. On the other hand, the selectivity showed a maximum at a silver crystallite size of about 10 nm, and this occurred well before the final deactivated level was attained. Processes such as this which demonstrate the effect of crystallite size and dispersion on activity and selectivity have been named "structure sensitive".

A further class of sintering of metallic surfaces is that in which the chemisorption of a poison on the metal surface promotes crystallite growth. This process might be supposed to occur by a combination of altered thermal and geometric effects on the surface caused by this chemisorption. An example of this behaviour, provided by Clay and Petersen (1970), is when arsine is adsorbed on platinum. It should also be pointed out that these processes may be enhanced by the exothermic adsorption of the poison on the crystallites.

References

Appleby, W. G., Gibson, J. W. and Good, G. M. (1962). *Ind. Eng. Chem.* (*Proc. Des. Devel.*) **1**, 102.
Bache, H. H. (1955). *J. Amer. Ceram. Soc.* **53**, 1205.
Baker, R. T. K. and Chludzinski, J. J. (1980). *J. Catal.* **64**, 464.
Blue, R. W. and Engle, C. J. (1951). *Ind. Eng. Chem.* **43**, 494.
Bridger, G. W. (1970). In *Catalyst Handbook*. Wolfe Scientific Books.
Brunn, L. W., Montagna, A. A. and Paraskos, J. A. (1976). *Amer. Chem. Soc. Prepr.* **21**, 173.
Butt, J. B., Delgado-Diaz, S. and Muno, W. E. (1975). *J. Catal.* **37**, 158.
Ciapetta, F. G., Dobres, R. M. and Baker, R. W. (1958). *J. Catal.* **6**, 495.
Clay, R. D. and Petersen, E. E. (1970). *J. Catal.* **16**, 32.
Coble, R. L. J. (1967). *Appl. Phys.* **32**, 79.
Eberly, P. E., Kimberlin, C. N., Miller, W. H. and Drushel, H. V. (1966). *Ind. Eng. Chem.* (*Proc. Des. Devel.*) **5**, 193.
Flynn, P. E. and Wanke, S. E. (1975). *Catal. Rev. Sci. Eng.* **12**, 93.
Fowle, M. J., Bent, R. D., Ciapetta, F. G., Pitts, P. M. and Leum, L. N. (1951). *Adv. Chem. Ser.* **5**, 76.
Frayer, J. W., Montagna, A. A. and Yanik, S. J. (1975). Jap. Petrol. Inst., Fuel Oil Desulph. Symp. (Tokyo), May.
Goldstein, M. S. and Morgan, T. R. (1970). *J. Catal.* **16**, 232.
Gruher, H. L. (1962). *J. Phys. Chem.* **66**, 48.
Haldeman, R. G. and Botty, M. C. (1959). *J. Phys. Chem.* **63**, 489.
Hastings, K. E., James, L. C. and Mounce, W. R. (1975). *Oil Gas J.* **73**(26), 122.
Hermann, R. A., Adler, S. F., Goldstein, M. S. and DeBaun, R. H. (1961). *J. Phys. Chem.* **65**, 2189.
Hughes, R. and Zadeh, H. M. (1976). *Proc. 1st. Europ. Symp. on Thermal Analysis* (ed. D. Dollimore), p. 131. Heyden, London.
Hughes, T. R., Houston, R. J. and Sieg, R. P. (1962). *Ind. Eng. Chem.* (*Proc. Des. Devel.*) **1**, 96.
Inoguchi, M. (1976). *Shokubai* **18**(3), 78.
Johnson, D. L. (1969). *J. Appl. Phys.* **40**, 192.
Kuczynski, G. X., Hooten, N. A. and Bibbon, C. F. (eds) (1967). *Sintering and Related Phenomena*. Gordon and Breach, New York.
Kwan, T. and Sato, M. (1970). *Nippon Kagaku Kaishi* **91**, 1103.
Levinter, M. E., Ponchenkov, G. M. and Tanatarov, M. A. (1967). *Int. Chem. Eng.* **7**, 23.
Luss, D. (1970). *Chem. Eng. J.* **1**, 311.
Maat, H. J. and Moscou, L. (1956). *Proc. 3rd Int. Congr. Catalysis*, p. 1277. North-Holland.
Massoth, F. E. (1967). *Ind. Eng. Chem.* (*Proc. Des. Devel.*) **6**, 200.
Maxted, E. B. (1951). *Adv. Catal.* **3**, 129.

Maxted, E. B. and Marsden, A. (1946). *J. Chem. Soc.* 766.
McIver, D. S., Tobin, H. H. and Barth, R. T. (1963). *J. Catal.* **2**, 485.
Mills, G. A., Boedeker, E. R. and Oblad, A. G. (1950). *J. Amer. Chem. Soc.* **72**, 1554.
Montagna, A. A., Chun, S. W. and Frayer, J. A. (1975). 9th World Petrol. Congr. **PD18**, 1.
Moritz, K. H. *et al.* (1970). Jap. Petrol. Inst., Fuel Oil Desulph. Symp. (Tokyo), May 10–11th.
Newson, E. J. (1970). Preprints A141–152. Div. Petrol. Chem., 160th National Meeting Amer. Chem. Soc., Chicago.
Newson, E. J. (1975). *Ind. Eng. Chem.* (*Proc. Des. Devel.*) **14**, 27.
Ohtsuka, T. (1977). *Catal. Rev. Sci. Eng.* **16**, 291.
Oxenreiter, M. F., Frye, C. G., Hoekstra, G. B. and Sroka, J. M. (1972). Jap. Petrol. Inst., Fuel Oil Desulph. Symp. (Tokyo).
Ozaki, H., Satomi, Y. and Hisamitsu, T. (1975). 9th World Petrol. Congr. (Tokyo) **PD18**.
Ozawa, Y. and Bischoff, K. B. (1968). *Ind. Eng. Chem.* (*Proc. Des. Devel.*) **7**, 67.
Paal, C. and Hartman, W. (1918). *Ber.* **51**, 711, 894.
Paal, C. and Windisch, E. (1913). *Ber.* **46**, 4010.
Parera, J. M. (1968). *J. Res. Inst. Hokkaido Univ.* **16**, 525.
Pines, H. and Haag, W. O. (1960). *J. Amer. Chem. Soc.* **82**, 2471.
Presland, A. E. B., Price, G. L. and Trimm, D. L. (1972). *J. Catal.* **26**, 313.
Ramser, J. N. and Hill, P. B. (1958). *Ind. Eng. Chem.* **50**, 117.
Ristic, M. M. (1969). *Phys. Sinter.* **1**, A1–A14; **2**, 49.
Rudershausen, C. G. and Watson, C. C. (1955). *Chem. Eng. Sci.* **3**, 110.
Scott, J. W., Bridge, A. G., Christiansen, R. J. and Gould, G. D. (1970). Jap. Petrol. Inst., Fuel Oil Desulph. Symp. (Tokyo), March 10–11th.
Tohdoh, N. *et al.* (1971). *Kogyo Kagakushi.* **74**, 563.
Trimm, D. L. (1977). *Catal. Rev. Sci. Eng.* **16**, 135.
Voorhies, A. (1945). *Ind. Eng. Chem.* **37**, 318.
Webb, G. and McNab, J. I. (1972). *J. Catal.* **26**, 226.
Williams, A., Butler, G. A. and Hammonds, J. (1972). *J. Catal.* **24**, 352.
Wilson, J. L. and Den Herder, M. J. (1958). *Ind. Eng. Chem.* **50**, 305.

3
Diffusion and Deactivation of Catalysts

3.1 Introduction

In the preceding chapters it has been assumed implicitly that any impurity poison in the feed will have free access to all points of a catalyst surface with no restriction. Similarly, any fouling reaction has been assumed to occur with no physical hindrance on all points of the catalyst surface. These assumptions would be valid for all circumstances if the catalyst was in the form of a very fine powder and the gas velocity over the surface was sufficiently high to make any external mass transfer resistance negligible. In practice, however, because of pressure drop considerations in a fixed bed reactor, it is not feasible to have the catalyst in the form of a fine powder and the catalyst is usually in the form of pellets and extrudates. Only in fluidized bed or slurry reactors are powdered catalysts employed, and even for these systems there may be certain physical restrictions that limit the theoretically attainable chemical rate.

Catalyst pellets and extrudates are usually made from a catalyst powder by a process of agglomeration from the original precipitated catalyst powder. The original precipitated particle is thus retained as a unit in the final pellet. The powder particle will possess an intrinsic pore structure of micropores. When particles are agglomerated to form a pellet the voids between the particles will give rise to a macropore structure. The situation is illustrated in Fig. 3.1 and it can be seen that the pellet has a bidisperse pore structure, i.e. it contains both micropores and macropores. Since the macropores are a consequence of the interparticle voidage it would be expected that the average diameter of these would be of the order of the dimensions of the particles themselves, i.e. about 1–10 µm. On the other hand the micropores are much smaller and typically may have pore diameters in the range 5–20 nm. It is the existence of this composite pore structure with very different groupings of pore diameters that may give rise to different diffusional mechanisms and therefore complicate any method of estimating effective transport of the gaseous species within the pores of the catalyst.

It should also be mentioned that some catalysts have only a monodisperse

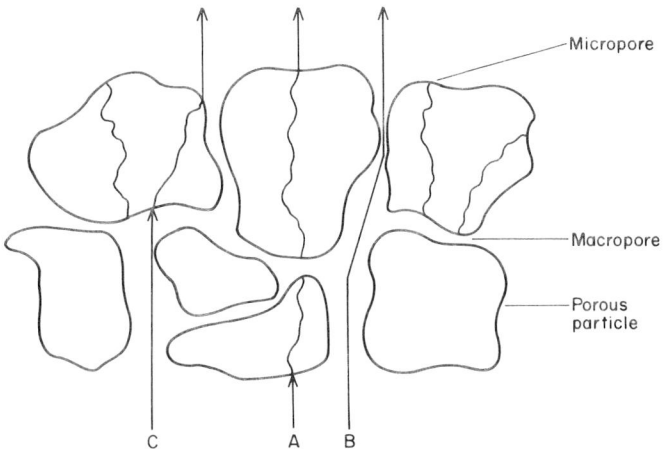

Fig. 3.1 Structure of porous catalyst pellet.

pore structure, i.e. there is only one predominant peak when pore size fraction is plotted against pore radius. Examples of such catalysts include the silica–alumina bead catalysts used in fixed and moving beds which are prepared by a gel technique.

Because catalytic gas–solid reactions involve the flow of gases past the solid materials there is always the possibility of a mass transfer limitation between the gas and the solid and an associated heat transfer limitation. Usually, in commercial practice, flow rates are sufficiently high to minimize any mass transfer resistance from this source but the effect can be important in some instances such as when relatively low gas velocities are used with large catalyst pellets.

Both internal diffusion and external mass transfer may act singly or combined to reduce the magnitude of the intrinsic chemical rate on the catalyst surface. A diagram showing the effect of the different resistances on the reactant concentration at various points outside and within the catalyst pellet is portrayed in Fig. 3.2. If the catalyst is very active the reactant concentration may drop to zero at some point within the pellet. The overall reaction may be assumed to consist of a number of steps:

(1) Mass transfer of reactants from the bulk gas to the external surface of the catalyst pellet.
(2) Diffusion of reactants along the pores of the pellet.
(3) Adsorption of reactant onto the active sites of the catalyst.
(4) Surface reaction on the active sites.
(5) Desorption of products from the active sites.

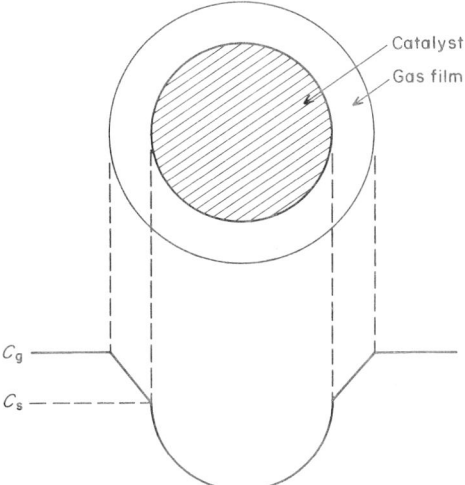

Fig. 3.2 External and internal concentration gradients in a reacting catalyst pellet.

(6) Diffusion of the products back along the pores to the outside surface of the pellet.
(7) Mass transfer of the products from the external surface of the catalyst pellet to the bulk gas phase.

Each of these steps will proceed at a certain rate, and if a steady state is attained the slowest will be the rate-determining step. Because chemical rates have a much larger temperature coefficient than rates of mass or heat transfer, the overall rate of a process tends to be controlled by the physical steps at high temperatures. The processes of mass and heat transfer outside and within the catalyst pellets will be considered next since they constitute an important factor in a detailed study of catalyst deactivation.

3.2 Mass and heat transport in catalyst pellets

3.2.1 External mass and heat transfer

In a heterogeneous catalytic gas–solid reaction, the reactants are transported from the bulk fluid to the outer surface of the catalyst and the reaction products transported back to the bulk fluid. If non-isothermal effects are important, the reaction heat is also transferred from the catalyst surface to the bulk fluid. The rate of mass and heat transfer between the fluid and the outside surface of the catalyst is determined primarily by the nature of the fluid flow through the reactor. When the fluid passes over the surface of a catalyst a boundary layer develops, in which the velocity of the fluid varies from zero at the particle surface to the bulk stream velocity at the boundary. Transport of reactants and products across this laminar gas film is by bulk diffusion, and the heat produced or required by the reaction is transported by an analogous mechanism.

The external mass and heat transfer steps have been extensively studied, mainly through the examination of simple non-reactive physical processes. Thus, in the case of mass transfer, vaporization and drying techniques have been used to establish correlations in terms of an empirical constant, namely the mass transfer coefficient. This is a constant connecting the flux of a particular molecular species to the concentration difference between the bulk fluid and the external surface:

$$N = k_c(C_o - C_s) \tag{3.1}$$

where k_c is the mass transfer coefficient and N is the molar flux.

Similarly, the heat transfer flux from the outer pellet surface to the bulk fluid per unit external area of catalyst is:

$$Q = h(T_s - T_o) \tag{3.2}$$

Mass transfer coefficients are usually correlated by means of the Sherwood number, Sh, while heat transfer coefficients are similarly correlated by the Nusselt number, Nu. These are defined as:

$$\text{Sh} = \frac{k_c d_p}{D_{AB}} \qquad \text{Nu} = \frac{h d_p}{K} \tag{3.3}$$

where d_p is the pellet diameter and D and K are the gas diffusivity and thermal conductivity respectively.

Dimensional analysis has shown that the Sherwood number should be a function of the Reynolds number, Re, and the Schmidt number, Sc, while the Nusselt number should be a function of both the Reynolds number and the Prandtl number, Pr. There is a vast literature giving specific form to these general formulae and it is beyond the scope of this monograph to review these fully. However, mention should be made of some of the more useful of these. One early correlation was that established by Ranz and Marshall (1952):

$$\text{Sh} = 2.0 + 0.6 \, \text{Re}^{1/2} \, \text{Sc}^{1/3} \tag{3.4}$$

This relation was found to be valid for particle Reynolds numbers from 0 to 200. Frössling (1938) had shown that the Sherwood number should have a value of 2 in a stagnant fluid and that there were theoretical agreements supporting a square-root dependence on Reynolds number. Rowe and

Claxton (1965) reviewed earlier work and, in experiments on heat and mass transfer from spheres to air and water, developed correlations for the extended Reynolds number range from 20 to 2000. Equations similar to (3.4) were obtained for both mass and heat transfer, but with values of the constant slightly different from that of 0.6 quoted by Ranz and Marshall.

Classically, mass and heat transfer coefficients have been obtained from the j_D and j_H factors for mass and heat transfer respectively, first proposed by Chilton and Colburn (1934). Using this approach the j_D factor is defined by:

$$j_D = \frac{\text{Sh}}{\text{Re Sc}^{1/3}} \tag{3.5}$$

Appropriate substitution of Nu for Sh and of Pr for Sc gives the corresponding relation for j_H.

Because of their semi-theoretical basis, correlations of the Ranz–Marshall type are preferred and will be used generally in this text.

3.2.2 Intrapellet mass transfer

Three basic types of diffusion may be identified in porous catalyst pellets. The actual mode of transport within the porous structure will depend on the pore radius and the pressure conditions within the reactor. The transport of gases within the pores may occur by (a) bulk diffusion, (b) Knudsen diffusion, and (c) surface diffusion. In many instances diffusion is by a combination of bulk and Knudsen diffusion; this is known as the transition region.

If the mean free path of the diffusing molecules is small compared with the pore diameter, collisions between molecules control the diffusion process. This type of diffusion is known as bulk or molecular diffusion. For equimolar counterdiffusion of species A diffusing in a mixture of species A and B, the molar flux is proportional to the concentration gradient in the direction of diffusion:

$$N_A = -N_B = -D_{AB}\frac{dC_A}{dL} \tag{3.6}$$

where N_A and N_B are the molar fluxes of the species A and B, D_{AB} is the bulk diffusion coefficient for molecules of A in the mixture, and C_A is the concentration of A.

For a binary system D_{AB} may be estimated by the equation based on kinetic theory and the Lennard-Jones expression for intermolecular forces:

$$D_{AB} = \frac{0.001858 T^{3/2}[(M_A + M_B)/M_A M_B]^{\frac{1}{2}}}{P\sigma_{AB}^2 \Omega_D} \tag{3.7}$$

where M_A and M_B are the molecular weights of A and B, P is the total pressure, Ω_D is the collision integral, and σ_{AB} is the collision diameter of the two molecules. Values of the collision integral are obtained from tables of the force constant and are tabulated in standard texts such as that of Hirschfelder et al. (1954). Equation (3.7) shows that D_{AB} is proportional to $T^{3/2}$ and is inversely proportional to total pressure. It is, however, independent of pore size.

Knudsen diffusion occurs when the mean free path of the molecules is considerably greater than the pore diameter. Under these conditions collisions of molecules with the pore walls are more frequent than intermolecular gaseous collisions. Knudsen demonstrated that the net flow of molecules in the direction of gas flow is proportional to the gradient of the molecular flux. Using the proportionality factor $8\pi r_p^3/3$ results in a Knudsen diffusion coefficient:

$$D_{K_A} = \frac{8}{3} r_p \sqrt{\frac{RT}{2\pi M_A}} \tag{3.8}$$

in which r_p is the pore radius. This equation applies for straight round pores only, and since the majority of porous solids are not well represented by straight round pores the above equation cannot be applied directly. A straight round pore has a volume to surface ratio of $r_p/2$ and therefore the mean pore radius, r_e, may be defined as:

$$r_e = \frac{2V_b}{S_b} = \frac{2\varepsilon}{S_b \rho_b} \tag{3.9}$$

where the subscript b refers to the catalyst pellet and ε and ρ are the void fraction and density. The Knudsen equation thus becomes:

$$D_{K_A} = \frac{16}{3} \frac{\varepsilon}{\rho_b S_b} \sqrt{\frac{RT}{2\pi M_A}} \tag{3.10}$$

The Knudsen diffusion coefficient is independent of pressure but is directly proportional to the pore size and inversely proportional to the square root of the molecular weight of the diffusing species.

The transition region is the region where neither Knudsen nor bulk diffusion control predominates. The relationship between the flux and concentration gradient in this region for binary gas diffusion was obtained by Scott and Dullein (1962) and is:

$$N_A = -\frac{1}{\frac{1-\alpha Y_A}{D_{AB}} + \frac{1}{D_{K_A}}} \frac{dC_A}{dL} \tag{3.11}$$

where $\alpha = 1 + N_B/N_A$.

For equimolar counterdiffusion this equation reduces to the familiar form first given by Pollard and Present (1948):

$$N_A = -\frac{1}{\dfrac{1}{D_{AB}} + \dfrac{1}{D_{K_A}}} \frac{dC_A}{dL} \qquad (3.12)$$

Transport by movement of the molecules over the surface is known as surface diffusion and occurs in the direction of decreasing surface concentration. Since adsorption is a function of the gas pressure of the adsorbed species in the region adjacent to the surface, both surface diffusion and bulk diffusion act in the same direction. Although intraparticle transport often has an important and sometimes limiting influence on the overall rate of a heterogeneous chemical reaction, the particular circumstances that make surface transport a significant factor are less clear. Indeed, to date, there are only a few examples of situations where surface diffusion has been shown to be important.

The complicating factor in diffusion in catalyst pellets is the pore structure and the lack of detailed knowledge of this. The internal pore structure consists of pores of non-uniform cross section, which pursue a very tortuous path through the catalyst and intersect with many other pores. Thus, the diffusivity estimated for a cylindrical straight pore must be modified by a multiplying factor to account for the geometry of the pore. A number of models have been proposed to predict the effective diffusivity of a porous catalyst pellet. The optimum model is the one that gives a realistic representation of the geometry of the pores but can be described in terms of easily measurable properties of the catalyst pellets, such as the surface area and pore volume, the density of the solid phase, and the pore size distribution. Some commonly used models are summarized below.

Probably the simplest model for diffusion in a porous solid assumes that transport of the molecules occurs through a number of parallel, equal sized, capillaries (Wheeler, 1955). The flux is given by this model as:

$$N_{AS} = n_p \pi r_p^2 \frac{L}{L_e} N_A \qquad (3.13)$$

where N_{AS} is the superficial molar flux, N_A is the molar flux, in each capillary of radius r_p, and $n_p \pi r_p^2$ is the total pore area per unit solid area for n_p pores. The term L_e/L gives the ratio of the actual diffusion path length to the straight-line distance through the solid, on the basis of randomly orientated pores with an average angle of 45° to the surface, Wheeler suggested that this should have the value of $\sqrt{2}$. The porosity of the solid, ε, equals $n_p \pi r_p^2 L_e/L$, so equation (3.13) becomes:

$$N_{AS} = \varepsilon \left(\frac{L}{L_e}\right)^2 N_A \tag{3.14}$$

or

$$N_{AS} = \frac{\varepsilon}{2} N_A \tag{3.14a}$$

using the substitution $L_e/L = \sqrt{2}$ above.

A more general expression arises if the irregular shape as well as the meandering nature of the pores is taken into account by means of a tortuosity factor, τ. Equation (3.14) now takes the form:

$$N_{AS} = \frac{\varepsilon}{\tau} N_A \tag{3.15}$$

or in terms of diffusion coefficients:

$$D_{eA} = \frac{\varepsilon}{\tau} D_{AB} \tag{3.16}$$

in which D_{eA} is the effective diffusion coefficient based on unit cross section of the porous solid. On the basis of Wheeler's model $\tau = 2$, as in equation (3.14a); this probably represents a lower limit to τ, which typically ranges from this value up to about 10. Equation (3.16) is written in terms of bulk diffusion. A similar expression should be written for Knudsen flow which allows for the effect of internal pore structure:

$$D_{eK} = \frac{\varepsilon}{\tau} D_K \tag{3.17}$$

This may then be combined with D_{eA} to get an overall effective diffusivity by employing the relation implicit in equation (3.12), that is:

$$\frac{1}{D_e} = \frac{1}{D_{eA}} + \frac{1}{D_{eK}} \tag{3.18}$$

The above model is only suitable for catalysts with a narrow pore size distribution unless ordinary bulk diffusion dominates so that the diffusion flux becomes independent of pore size. Johnson and Stewart (1965) modified this model to take into account the more usual broader pore distributions. A pore size distribution function $f(r)$ was defined such that $f(r)\,dr$ was the void fraction of pores with radii between r and $(r + dr)$. The flux was then calculated as:

$$N_{AS} = \frac{D_{AB} P \varepsilon}{\tau R T L \alpha} \int_0^\infty \ln\left[\frac{1 - \alpha Y_{AL} + D_{AB}/D_{KA}}{1 - \alpha Y_{A0} + D_{AB}/D_{KA}}\right] f(r)\,dr \tag{3.19}$$

where $\alpha = 1 + N_{BS}/N_{AS}$ and ε is the porosity. This model is usually termed the parallel pore model.

Wakao and Smith (1962) developed a model based on a bimodal pore size distribution and known as the random pore model. They suggested that the diffusion flux through a bimodal pore system is the sum of the contributions for the micropores, the macropores, and the micropores and macropores in series. The magnitude of each contribution depends on an effective cross sectional area for each. The resultant effective diffusivity based on this model may be written as:

$$D_{eA} = D_a \varepsilon_a^2 + \frac{\varepsilon_i^2(1 + 3\varepsilon_a)}{1 - \varepsilon_a} D_i \qquad (3.20)$$

where D_a and D_i are the diffusivities of macropores and micropores respectively. Application of this model requires a knowledge of the void volumes and mean pore radii for both macropore and micropore regions, but does not require any knowledge or assumptions of a tortuosity factor.

All the above has been written implicitly for binary systems with the assumption of equimolar counterdiffusion. The latter condition implies that the reaction considered is of the simple type A → B, while the limitation to binary systems is often restrictive for practical situations. In such cases the multicomponent diffusion flux equations of the Stefan–Maxwell type must be employed for the bulk diffusion coefficients (Knudsen diffusion coefficients are based on one component only, so this problem does not arise for this type of diffusion). In general, the diffusion coefficients so derived, D_{im}, are dependent on position. For many cases, where such dependence is quite small, the binary diffusion formulas may be generalized by replacing D_{AB} by D_{im}. This may be applied with only small error when one component is limiting in a gas mixture.

The modified Stefan–Maxwell equations are usually written in the form:

$$\frac{1}{cD_{im}} = \frac{\sum\limits_{j=1}^{n} (1/cD_{ij})(x_j N_i - x_i N_j)}{N_i - x_i \sum\limits_{j=1}^{n} N_j} \qquad (3.21)$$

Use of these relations is developed by Hirschfelder *et al.* (1954), while an excellent review of external and internal mass transfer processes in porous catalysts is given by Satterfield (1970).

3.2.3 Intraparticle heat transport

In a porous catalyst pellet, transport of heat may occur through the gas in the pores or through the solid. Transport through the solid is usually the major contribution to the total heat flux, and the effective thermal conductivity of a catalyst pellet (defined as the heat transferred per unit total area perpendicular to the direction of heat transfer) increases as the void fraction decreases. In a pellet composed of an aggregate of particles, this effect is most easily interpreted as due to the increased area of contact between particles as the void fraction decreases.

The effective thermal conductivity of porous catalysts depends mainly on the dimensions and distribution of the pores and the intrinsic solid thermal conductivity. For pellets made by compaction of particles, satisfactory agreement between predictions and measurements has been obtained using the concept first developed by Harriott (1975) in which the predictive method is applied in turn to both particles and pellets. This method has been shown to work well for oxide catalysts (Sharma *et al.*, 1975; Soomro and Hughes, 1979) even with quite simple models for the pore structure, but predictions for supported metal catalysts are less satisfactory.

3.3 Diffusion and reaction in porous catalysts

Because of the finite pellet size of most catalyst pellets used in fixed beds, a concentration gradient will develop within the porous catalyst, as described in Fig. 3.2. The decrease in reactant concentration within the pellet is caused by the combined effects of diffusion and reaction, and the net effect is (for isothermal pellets) to decrease the overall reaction rate for the pellet. Depending on the relative magnitudes of the diffusion rate and the chemical reaction rate, a molecule of reactant may penetrate deep into the pellet before reacting or may react close to the surface. In the former case the concentration profile will be relatively shallow compared with the latter where in extreme cases the concentration of reactant may drop to zero at a point quite close to the surface.

For an isothermal spherical pellet, a steady-state mass balance over a spherical shell of thickness dr yields the differential equation for diffusion and reaction:

$$D_e \frac{d^2 C}{dr^2} + D_e \frac{2}{r} \frac{dC}{dr} + R^*[C, T] = 0 \tag{3.22}$$

in which D_e, the effective diffusivity, is assumed constant and independent of radial position, and $R^*[C, T]$ is the true reaction rate expressed in general terms as a function of concentration and temperature.

The corresponding steady-state heat balance taken over the same spherical shell gives:

$$K_e \frac{d^2 T}{dr^2} + K_e \frac{2}{r} \frac{dT}{dr} - R^*[C, T](-\Delta H) = 0 \qquad (3.23)$$

The intraparticle diffusion limitation in porous catalyst pellets undergoing reaction was first recognized by Damköhler (1937), Thiele (1939), Zeldovitch (1939), and Wheeler (1951). For isothermal steady-state conditions, equation (3.22) may be solved analytically to obtain the concentration profile. From this the overall reaction rate in a pellet may be obtained. This then leads to the concept of catalyst utilization or, as it is more usually called, an effectiveness factor. The effectiveness factor, η, for a catalyst pellet is defined as:

$$\eta = \frac{\text{Observed reaction rate in pellet}}{\substack{\text{Reaction rate if all pellet surface corresponded to} \\ \text{concentrations and temperature in the bulk gas phase}}} \qquad (3.24)$$

The advantage of employing this concept is that reaction rates may be defined in the form:

$$R^* = f(C_o, T_o)\eta \qquad (3.25)$$

where C_o and T_o refer to the concentration and temperature in the bulk gas phase. If the effectiveness factor is known, considerable simplification in the design of catalytic reactors is achieved, since the effectiveness factor, η, automatically accounts for any diffusional limitations and makes it unnecessary to solve the differential equations for heat and mass transport for each pellet within the catalyst bed.

For isothermal conditions and normal concentration dependence of the reaction rate, η may be interpreted physically as the proportion of the pellet volume used in the reaction. In an isothermal system the effectiveness factor can have a maximum value of unity, and for a first-order reaction in a spherical catalyst pellet, with no external mass transfer resistance, solution of equation (3.22) using the definition (3.25) of η yields the following analytical solution for η:

$$\eta = \frac{3}{\phi}\left(\frac{1}{\tanh \phi} - \frac{1}{\phi}\right) \qquad (3.26)$$

where ϕ is the Thiele modulus and represents the ratio of the intraparticle diffusional resistance to the chemical reaction resistance. For a general nth order reaction, ϕ is defined by:

$$\phi = R\sqrt{\frac{kC_o^{n-1}}{D_e}} \qquad (3.27)$$

in which R is the radius of the catalyst pellet.

For a single pore or flat slab of catalyst the effectiveness factor for isothermal conditions, with no external mass transfer resistance, is given by an equation analogous to (3.26) but somewhat simpler:

$$\eta = \frac{\tanh \phi}{\phi} \qquad (3.28)$$

and in the Thiele modulus the pore length or slab thickness, L, replaces R in equation (3.27).

Usually the influence of ϕ on the effectiveness factor is portrayed in the form of logarithmic plots of η against ϕ; an example is given in Fig. 3.3 for isothermal reaction. Various pellet geometries and reaction orders can be shown similarly but the effect of these variations is not very great. Essentially, at low values of the Thiele modulus the effectiveness factor has a value close to unity. This is to be expected since a low value of ϕ implies a small pellet with a small value of the rate constant and/or a high value of the effective diffusivity. Under these conditions the reaction would be expected to be under chemical resistance control, with little or no influence of the diffusional resistance (large D_e and small pellet size), and hence the effectiveness should be around unity. At the other extreme, with large sized pellets, a high value for k, and only a small diffusivity value, the process

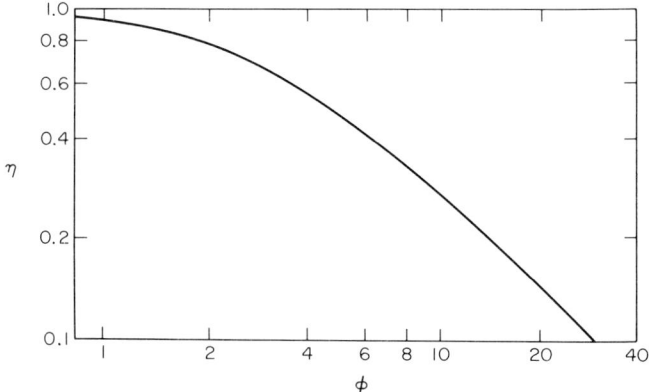

Fig. 3.3 Logarithmic plot of effectiveness factor (η) against Thiele modulus (ϕ) for an isothermal first order reaction in a sphere.

would be diffusion controlled and the value of η will rapidly decrease with increase of ϕ. This is shown in Fig. 3.3. This region where η is inversely proportional to ϕ is known as the asymptotic region and is frequently attained in reactor operation, since commercial catalysts usually have high activity and are often employed as relatively large pellets. In the asymptotic region, monodisperse catalyst pellets operating isothermally will give an apparent activation energy with a limiting value equal to half the true activation energy. For non-isothermal systems and those exhibiting a bidisperse pore distribution the situation is more complex; values of the apparent activation energy for these cases have been derived by Rajadhyaksha and Doraiswamy (1976).

Departures from the simple curves of Fig. 3.3 occur when non-isothermality is included in the analysis, and also when external mass and heat transport are important. It has been shown by several authors (e.g. Carberry, 1961; Weisz and Hicks, 1962) that, for an exothermic reaction, the effectiveness factor can have values greater than unity. This is due to the increased temperature rise in the pellet compensating for, and ultimately reversing, the effect of concentration decrease caused by the diffusional resistance of the pellet. Furthermore it was also shown that, at certain values of the Thiele modulus, more than one value of the effectiveness factor could be obtained. This multiple steady state region has important consequences in problems of reactor stability.

An important generalization first noted by Carberry (1966) is that for realistic values of the operating parameters the major heat transfer resistance is external to the catalyst pellet while for mass transfer the external resistance is usually small with the major resistance to mass transfer residing within the porous catalyst.

If all effects are included in plots of the effectiveness factor against Thiele modulus, these may be divided into four regions as depicted in Fig. 3.4. These are the kinetic region, the diffusion controlled region, the multiple steady state region, and the region of interphase (external) mass transfer. In the kinetic region where the Thiele modulus is small, the chemical reaction rate is controlling and the effectiveness factor is approximately unity. As the Thiele modulus increases, the intraphase diffusion becomes the controlling resistance and the effectiveness factor falls below unity. In this region the reaction is confined to the outer layers of the catalyst pellet, the interior being at a higher temperature than the surface but essentially isothermal. The major part of the overall temperature rise occurs across the gas film surrounding the catalyst particle. Further increase of the Thiele modulus yields region D where gas film mass transfer becomes dominant. The existence of a large interphase mass transfer resistance prevents the heat of reaction being removed sufficiently rapidly from the external catalyst surface. In this region

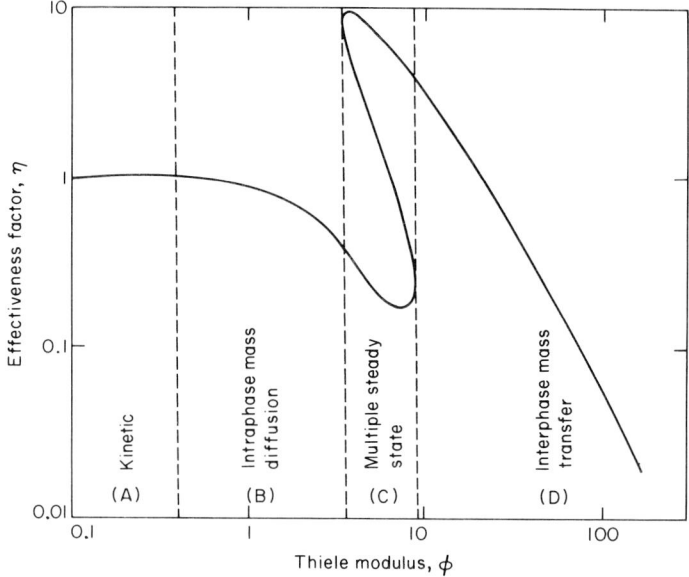

Fig. 3.4 Schematic representation of the main rate-controlling variables.

reaction is effectively occurring only on the external surface of the catalyst pellet, the interior being at the same temperature as the surface but appreciably higher than that of the surrounding gas. The transition from region B to region D is accompanied by a large increase in effectiveness factor in region C. This is a region of multiple steady states.

The incorporation of external transport resistances causes multiple steady states to occur at lower and more realistic values of the thermicity factor, β, defined as $\beta = (-\Delta H)D_e C_o / T_o K_e$. Thus in Fig. 3.4 a peak value of η equal to about 10 is achieved for a β value of 0.02. If external transport resistances were assumed to be absent, the value of β necessary to obtain this same peak effectiveness would be 0.4, which would only be an appropriate value for β in exceptional circumstances.

More detailed and extensive reviews of the problem of reaction coupled with mass and heat transfer resistances are given in a number of excellent texts to which the reader should refer. Space limitations restrict mention of all of these but particularly noteworthy are those of Aris (1975), Carberry (1976), and Satterfield (1970).

3.4 Deactivation and diffusion

3.4.1 Analogy between selectivity and deactivation mechanisms

Wheeler (1955) proposed three different types of selectivity mechanism to explain the kinetics of multiple reactions; he denoted these as types I, II, and III. A type I mechanism represents the case where two reactions take place independently on the same catalyst. Activity decay due to a feedstream impurity may be represented in a similar way:

$$A + S \to B + S$$
$$P + S \to P \cdot S$$

The poison, P, adsorbs on the active catalyst sites to yield a product P·S, thus reducing the number of active sites for reaction.

A type II mechanism is characterized by parallel reactions of a single reactant species. Here the reactant A is converted into either the desired product B or an undesired product C. The corresponding situation for activity decay is referred to in the literature as parallel fouling. In this case a portion of the reactant A is converted into a product A·S which remains attached to the internal surface of the catalyst:

$$A + S \to B + S$$
$$A + S \to A \cdot S$$

This type of deactivation represents reactant inhibition which would be associated with coke formation.

A type III mechanism is characterized by a reactant producing a desired intermediate product which in turn reacts further to produce an undesired product. The corresponding situation for activity decay is referred to in the literature as serious fouling, where the product is adsorbed on the catalyst surface:

$$A + S \to B + S$$
$$B + S \to B \cdot S$$

The product B adsorbs on the catalyst active sites and inhibits further reaction on these sites. This type of decay can also be associated with coke formation.

3.4.2 Catalyst deactivation

When the surface of a catalyst becomes deactivated by poisoning or fouling, further reaction is inhibited by virtue of a foreign molecule being adsorbed in the porous structure of the catalyst and covering a fraction of its active surface. The reactant must then be transported to the non-deactivated parts of the catalyst before reaction can occur; thus deactivation increases the average distance over which a reactant must diffuse through the porous structure. Two types of deactivation may be distinguished, uniform or non-uniform. Uniform or homogeneous deactivation may occur if the diffusion rate of the poison or fouling material is much slower than that of the main reaction process. It can be expected that the deactivating species will travel deep into the particle, yielding a catalyst with deactivation evenly distributed.

Non-uniform deactivation, termed "selective" poisoning by Wheeler (1955), is by definition the process where the activity drop is a non-linear function of the uncovered active sites. This means that the deactivating species are not uniformly distributed inside the catalyst, and hence a concentration gradient of these is built up inside the catalyst. These two extremes will be considered in turn but limiting the discussion to first order reactions for clarity.

For the case of homogeneous distribution of the deactivating species along the length of a catalyst pore with a fraction q of the pore surface deactivated, the simplest assumption to make is that the intrinsic activity k of the catalyst pore wall decreases to $k(1 - q)$. This gives a linear decrease of activity with the fractional area deactivated. However, owing to the diffusional restriction of rate in the pore, the activity is not given by $k(1 - q)$ but must involve the Thiele modulus, ϕ. From the definition of the effectiveness factor in equation (3.24) and the expression for η for a single pore in equation (3.28) it can be seen that the overall observed rate of reaction in a single catalyst pore may be expressed as:

$$\text{Observed rate} \propto \phi \tanh \phi \quad (3.29)$$

The Thiele modulus for the deactivating catalyst is obtained by substituting $k(1 - q)$ for the rate constant in the equation defining this modulus, i.e.

$$\phi_d = \phi\sqrt{1 - q} \quad (3.30)$$

The ratio of the rates in the deactivated and non-deactivated processes is then:

$$\frac{R_d^*}{R^*} = \frac{\phi\sqrt{1 - q} \tanh(\phi\sqrt{1 - q})}{\phi \tanh \phi} = \frac{\sqrt{1 - q} \tanh(\phi\sqrt{1 - q})}{\tanh \phi} \quad (3.31)$$

For small ϕ (small diffusion effect), $R_d^*/R^* \to (1 - q)$, and the intrinsic non-selective nature of the poisoning is retained. However, when transport limitations are large, the tanh terms approach unity and $R_d^*/R^* \to \sqrt{1 - q}$. Wheeler termed this "anti-selective" deactivation since the activity falls off

3 Diffusion and Deactivation of Catalysts

less than linearly with concentration of deactivating species. Physically this occurs because, with a homogeneous distribution of poison, that fraction at the centre of the pellet will not influence the main reaction, which at high values of ϕ occurs mainly in the outer layers of the catalyst pellet. Fig. 3.5 shows a plot of fraction of activity remaining against the fraction of surface deactivated. Curve B depicts the linear decline in activity with fraction deactivated, which occurs when ϕ is very small (i.e. $\eta \approx 1$), while curve A is for anti-selective deactivation (ϕ very large).

Curves C and D in Fig. 3.5 refer to "pore mouth" deactivation. This situation occurs when the affinity of the catalyst for the poison or deactivating species is so great that, when a small concentration of poison is present, the outer pore mouths become completely poisoned while the inner pore structure is still active. The situation under these conditions may be represented pictorially as in Fig. 3.6. If a pore of length L has sufficient poison added to cover a fraction q of its surface, the length qL nearest to the pore mouth will be completely poisoned and a length $(1 - q)L$ beyond this will retain the original catalyst activity. Thus molecules must diffuse through a length qL before reaching the active surface.

From Fig. 3.6, C_0 is the concentration of reactant at the pore mouth, while C_L is the concentration at the point qL from the pore mouth where the deactivated zone ends. At steady state the diffusion flux through this deactivated zone must equal the rate of reaction in the active region of length $(1 - q)L$. The concentration gradient in the deactivated outer layer will be linear, i.e. $\Delta C/\Delta L = (C_0 - C_L)/qL$, and the reaction rate in the length

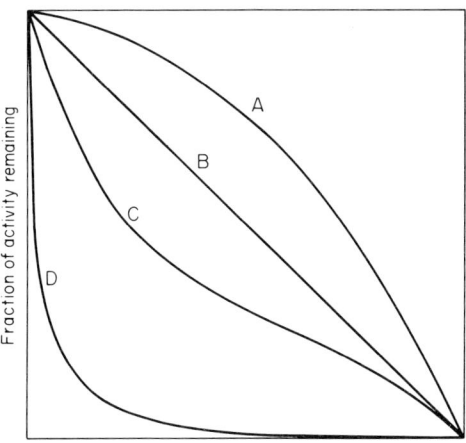

Fig. 3.5 Representation of different types of deactivation in a porous catalyst pellet. (Wheeler, 1955)

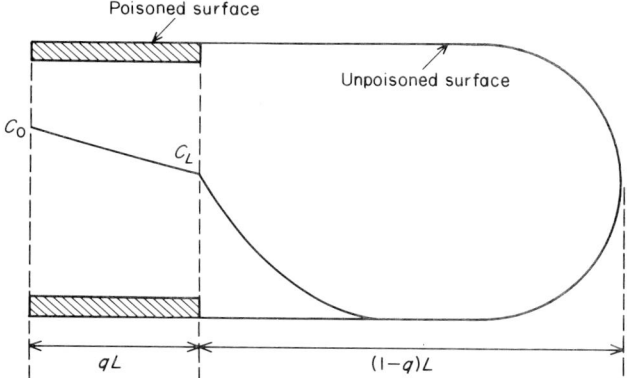

Fig. 3.6 Schematic of deactivated catalyst pore.

$(1 - q)L$ is given by the product of the diffusivity in the pore and the concentration gradient at $x = qL$, i.e.

$$R^* = -D_e \frac{dC_L}{dx}\bigg|_{x=qL} \quad (3.32)$$

For a cylindrical pore the concentration gradient at $x = qL$ is given by:

$$-\frac{dC_L}{dx} = \phi \frac{C_L}{L} \tanh \phi \quad (3.33)$$

and therefore, equating the fluxes between the poisoned and active zones, we have:

$$D_e \frac{C_0 - C_L}{qL} = \frac{D_e}{L} C_L \phi \tanh[\phi(1 - q)] \quad (3.34)$$

Solving the above equation explicitly for C_L gives:

$$C_L = \frac{C_0}{1 + \phi q \tanh[\phi(1 - q)]} \quad (3.35)$$

Hence, the rate of reaction in the partly poisoned pore is:

$$D_e \frac{C_0 - C_L}{L} = \frac{C_0 D_e}{L} \frac{\phi \tanh[\phi(1 - q)]}{1 + \phi q \tanh[\phi(1 - q)]} \quad (3.36)$$

In an unpoisoned pore the reaction rate is $C_0 D_e \phi (\tanh \phi)/L$ as in equation (3.32) above, and so:

$$\frac{R_d^*}{R^*} = \frac{\tanh[\phi(1 - q)]}{1 + \phi q \tanh[\phi(1 - q)]} \frac{1}{\tanh \phi} \quad (3.37)$$

3 Diffusion and Deactivation of Catalysts

This relationship is also shown in Fig. 3.5 in curves C and D for two values of the Thiele modulus. Marked selectivity is shown, even though the portion of the catalyst that is deactivated is deactivated uniformly (i.e. completely). The reason for such extreme effects as those shown for the strongly diffusion limited example of curve D is that the outer layer of the catalyst particle which would normally constitute the surface area used for reaction is precisely the surface that is rendered inactive by the pore mouth deactivation.

The influence of deactivation on the change in observed rate with temperature for various values of the Thiele modulus was also considered by Wheeler (1951) and the results obtained are illustrated in Fig. 3.7. Here the ratio of the rate at a temperature T to that at 200 K is plotted against the

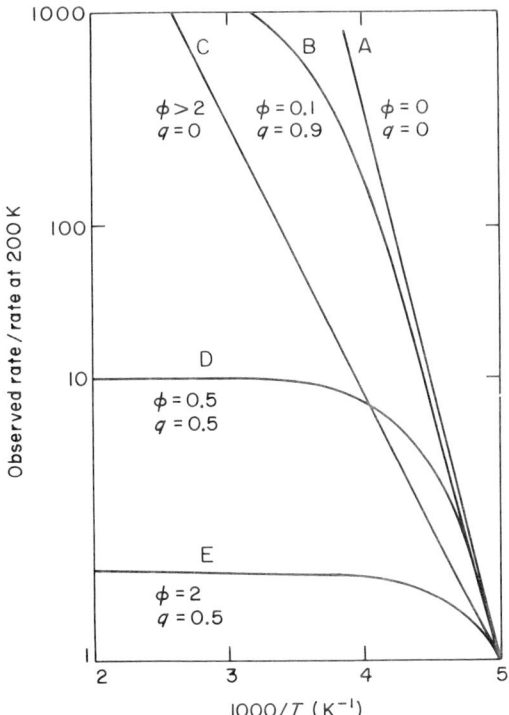

Fig. 3.7 Effect of poison and pore size on apparent activation energy. Plots of observed reaction rate against $1/T$ for a hypothetical catalyst having 46 kJ intrinsic activation energy (e.g. nickel in ethylene hydrogenation) but prepared with different pore sizes and poisoned to varying extent with poison preferentially adsorbed near the pore mouth: (A) large pores, no poison; (B) fairly large pores, 90% poisoned; (C) small pores, no poison; (D) moderate size pores, 50% poisoned. The horizontal portions of D and E correspond to diffusion-controlled reaction. (Wheeler, 1955)

reciprocal temperature for a reaction having an activation energy of 46 kJ mol^{-1}. The parameters identifying each curve are the Thiele modulus, ϕ, and the fractional coverage of the surface with poison, q. The values of ϕ quoted on the curves are those at 200 K; with increase in temperature the ϕ values will, of course, increase considerably. Curve A demonstrates a reaction unaffected either by diffusional restrictions or by deactivation ($\phi = 0, q = 0$) and would be appropriate for a catalyst pellet with very large pores. Curve C is for a catalyst containing small pores but again no deactivating species or poisons are present, so a straight line is obtained once more, but now with a decreased slope, demonstrating the well known apparent decrease in activation energy caused by diffusional resistances. Curve B is for large pores within the catalyst but with severe deactivation ($q = 0.9$). Under these conditions as the temperature is increased the path followed is very close to the non-deactivated case A, but at very high temperatures the reaction rate ratio decreases. This is due to the combined effect of (a) the onset of diffusion control caused by fast chemical reaction at these high temperatures, and (b) a decreasing portion of the non-deactivated surface area being available for reaction. Curves D and E represent extreme cases where, when an increase of temperature causes the reaction rate to become very high, the reaction will be controlled completely by the rate of diffusion through the deactivated pore mouth and the overall temperature coefficient is close to zero. Curve D is for moderate sized pores, while E represents a catalyst with small pores, and both D and E have q values equal to 0.5.

The real value of Wheeler's work was in quantifying the extremes of uniform deactivation and pore mouth deactivation. It should be emphasized that the analysis given above assumes that the amount and/or distribution of poison does not vary with time. In fact Wheeler's treatment considers only those circumstances in which a given amount of deactivating species is introduced into the system for a time sufficient to cause a fixed, and thereafter time independent, deactivation. More commonly, of course, a poison or coke depositing agent is fed to the catalytic system in a continuous stream, albeit at a very low concentration (in the case of an impurity poison). More commonly, therefore, the time dependent nature of the process should be considered in a more representative analysis. Also, Wheeler did not investigate the effect of external mass transfer resistance on catalyst deactivation. Both of these factors were taken into account by Carberry and Gorring (1966). In their paper, the fraction of deactivated catalyst was plotted against time for different values of the poison or deactivating species distribution. Both pore mouth and uniform distributions of poison were considered in the analysis which was also important in drawing a parallel between the shrinking core model for non-catalytic gas–solid reactions and deactivation processes by pore mouth poisoning. In both instances a zone travels to the centre of the particle with increasing time, and the reactant species have to diffuse through this zone before reaction can occur.

Carberry and Gorring observed that, with no external mass transfer restriction, pore mouth poisoning, up to about 40% of the total volume, showed a square root dependence on the time on stream. This type of relationship is often taken as representative of diffusion control. However, it was demonstrated that other combinations of deactivation distribution and mass transfer limitations could also lead to a similar time dependence, so observations of this kind do not lead to proof that a given mechanism is controlling.

3.5 Correlations for activity decay

Quantitative (and sometimes empirical) correlations between the amount of deactivation and catalytic activity have been reported in the literature over a considerable number of years. One of the earliest reported studies of this kind was that of Pease and Stewart (1925), but the first attempt at systematically classifying these was made by Maxted (1951) and his various coworkers. In most cases deactivation has been depicted by plotting the catalyst activity against the amount of poison present in the gas phase, since Maxted was mainly concerned with investigations into the effect of impurity poisons. Fig. 3.8 shows the decay of activity of a platinum catalyst used in the liquid-phase

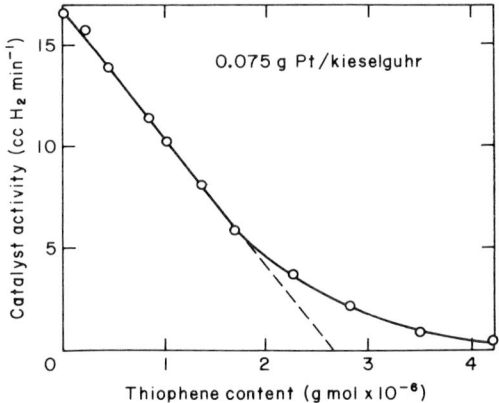

Fig. 3.8 Form of a typical poisoning curve, showing the depression of the activity of a supported platinum catalyst by increasing amounts of thiophene, in the liquid-phase hydrogenation of crotonic acid. (Maxted, 1951)

hydrogenation of crotonic acid, by increasing amounts of thiophene. The activity of the catalyst is seen to fall linearly with increasing poison content up to a point at which a substantial part of the catalyst activity is suppressed. It is then followed by a stage in which the activity of the catalyst falls far less steeply with further increase in poison content. Similar results to those shown in Fig. 3.8 were obtained in several other reactions by Maxted (1945). The linear portion of the plot can be represented by a relation of the form:

$$S = S_o - b_1 C_p \tag{3.38}$$

where S_o is the initial activity, b_1 is a poisoning coefficient, and C_p is the concentration of poison.

The linear fall in activity predicted by equation (3.38) is indicative of a homogeneous or non-specific type of deactivation, and may be represented as a simple decrease in the rate constant of the overall reaction, given by $k = k_o(1 - q)$ as used in Section 3.4.2. For poisoning by this "non-selective" mechanism, values of the poisoning coefficient b_1 may be obtained simply by comparing the slopes of plots of activity against poison content for various poisons. In many cases such plots can be linearized over a very small range, and therefore other functions have been suggested to account for the experimental results observed. These include the following types of expression:

$$S = S_o \exp(-b_2 C_p) \tag{3.39}$$

$$1/S = 1/S_o + b_3 C_p \tag{3.40}$$

$$S = b_4 C_p^d \tag{3.41}$$

These non-linear expressions were suggested as being representative of selective poisoning. If it is assumed that the amount of poison adsorbed is directly related to the time on stream, equations (3.38)–(3.41) may be derived from simple integration of differential activity *vs.* time expressions. Thus, the exponential form of equation (3.39) is based on the assumption of a first order dependence of the rate of activity decrease on the instantaneous activity of the catalyst. The various expressions in both differential and integrated form are summarized in Table 3.1.

The correlations in Table 3.1 are empirically defined through the observed activity and concentration of the poison present. Szepe and Levenspiel (1970) proposed a general power law:

$$-\frac{dS}{dt} = KS^d \tag{3.42}$$

which on integration corresponds to the expressions in Table 3.1.

Table 3.1 Correlations for catalyst deactivation.

Form of activity decay	Authors	System	Differential form of equation	Integrated form of equation	Exponent in power law
Linear	Maxted (1951)	Liquid phase hydrogenation of crotonic acid on Pt poisoned by AsH_3	$-\dfrac{dS}{dt} = b_1$	$S = S_o - b_1 t$	0
	Eley and Rideal (1941)	Para-H_2 conversion on W poisoned by O_2			
Exponential	Pease and Stewart (1925)	Hydrogenation of C_2H_4 on Cu poisoned by CO	$-\dfrac{dS}{dt} = b_2 S$	$S = S_o \, e^{-b_2 t}$	1
	Herrington and Rideal (1945)	Dehydrogenation of paraffins on chromia–alumina			
Hyperbolic	Germain and Maurel (1958)	Dehydrogenation of cyclohexane on Pt-Alumina	$-\dfrac{dS}{dt} = b_3 S^2$	$S^{-1} = S_o^{-1} + b_3 t$	2
Reciprocal power function	Voorhies (1954)	Catalytic cracking	$-\dfrac{dS}{dt} = b_5 A t^{-(b_5 + 1/b_5)}$	$S = A t^{-b_5}$	$b_5 + \dfrac{1}{b_5}$

3.6 Separable and non-separable kinetics

Strictly speaking equation (3.42), having no concentration term for any deactivating species, should apply only to deactivation where no concentration term is involved, i.e. deactivation caused by purely thermal sintering. Unfortunately it has been used for a wide range of deactivating phenomena and the conclusions obtained are therefore sometimes suspect when used in this way.

The simplest, but probably still the most satisfactory, form of expressing the overall rate of reaction under conditions where catalyst deactivation is important is as the product of two terms. The first of these depends solely on present conditions, while the second depends on past history. The overall rate may then be represented as:

$$R_d^* = R^*(\text{present conditions}) \cdot f(\text{past history}) \qquad (3.43)$$

It is important to recognize the limitations of separable kinetics. One important aspect is that separable kinetics cannot explain changes in selectivity, which arise as a consequence of the deactivation process. Similarly, preferential deactivation of one function of a bifunctional catalyst will not be accounted for by this model. These limitations arise because of the nature of equation (3.43) which corresponds to an ideal surface, characterized by uniform non-interacting sites. Because of the latter requirement, a chemisorbed molecule will not modify the properties of neighbouring free or occupied sites. However, it is well known that non-uniform sites and interactions between adsorbed molecules are the rule rather than the exception in catalytic systems. Such evidence includes, for example, changes in the heat of adsorption with coverage, the changes in selectivity due to poisoning (mentioned above), and the distribution of site strengths of acidic catalysts obtained by titration techniques (Chapter 2). The separable model has been used extensively for engineering applications despite these limitations and, although Gavalas (1971) has pointed out that it cannot predict the possibility of flow directional effects in catalytic reactors caused by non-separable kinetics, its use in general terms seems justified until more information is available on the specific nature of surfaces for a given reaction.

The current situation has been assessed very fully by Butt and coworkers quite recently (1978). They have shown that the concept of separability has on the whole been justified especially for fouling reactions. Most of the work referred to in this context has incorporated separable deactivation models into larger scale catalytic reactor models and fairly complex reaction kinetics, so any inconsistencies may well be masked by the extensive parameterization required. Perhaps it is more significant that most models of fouling by coking assume that the same surface sites are required for both the main reaction and the fouling reaction. Experimental evidence seems to justify this assumption. Similarly, for impurity poisoning, if the surface has a uniform activity for both the poisoning reaction and the main reaction, the use of separable kinetics will still be justified.

Only in the case of non-uniformity of surface sites for a catalyst deactivated by impurity poisons are difficulties with separable kinetics likely to arise. Examples include the work of Bakshi and Gavalas (1975) on the dehydration of lower alcohols over a silica–alumina catalyst poisoned by n-butylamine, and of Butt and coworkers (Weng et al., 1975) on the poisoning of the hydrogenation of benzene by thiophene over a supported nickel catalyst. In these last instances it would seem more appropriate to use a non-separable form of deactivation expression.

Therefore the situation may be summarized by noting that for deactivation by fouling or sintering the use of separable kinetics appears to be justified by experimental evidence. This is also true of impurity poisoning if the active sites of the catalyst are of uniform activity for both the main and the poisoning reaction, and there is no interaction between the adsorbed molecules of either reactant or poison. In other words, separable kinetics might be expected to hold for the ideal, Langmuir, type of adsorption since these assumptions are made if Langmuir adsorption holds. Furthermore, since many kinetic expressions for catalytic reactions are based on Langmuir adsorption, it would seem that, to be consistent, separable kinetics should be employed for these reactions. Only if detailed information is available on the non-uniformity of surfaces, or if the experimental data do not fit the separable form, should the non-separable form of kinetics be employed. Therefore, in the following treatment of the individual deactivation mechanisms, separable kinetics will be utilized.

References

Aris, R. (1975). *The Mathematical Theory of Diffusion and Reaction in Permeable Catalysts*. Clarendon Press, Oxford.
Bakshi, K. R. and Gavalas, G. R. (1975). *A. I. Ch. E. J.* **21**, 494.
Butt, J. B., Wachter, C. K. and Billimoria, R. M. (1978). *Chem. Eng. Sci.* **33**, 1321.
Carberry, J. J. (1961). *A. I. Ch. E. J.* **7**, 350.
Carberry, J. J. (1966). *Ind. Eng. Chem.* **58**, 40.
Carberry, J. J. (1976). *Chemical and Catalytic Reaction Engineering*. McGraw-Hill, New York and London.
Carberry, J. J. and Gorring, R. L. (1966). *J. Catal.* **5**, 529.
Chilton, T. H. and Colburn, A. P. (1934). *Ind. Eng. Chem.* **26**, 1183.
Damköhler, G. (1937). *Chem.-Ing.* **3**, 430. (1943). *Z. Phys. Chem.* **A193**, 16.
Eley, D. D. and Rideal, E. K. (1941). *Proc. Roy. Soc.* **A178**, 429.
Frössling, M. (1938). *Beitr. Geophys.* **52**, 170.

Gavalas, G. R. (1971). *Ind. Eng. Chem. (Fund.)* **10**, 621.
Germain, J. E. and Maurel, R. (1958). *Compt. Rend.* **247**, 1854.
Harriott, P. (1975). *Chem. Eng. J.* **10**, 65.
Herrington, E. F. G. and Rideal, E. K. (1945). *Proc. Roy. Soc.* **A184**, 434.
Hirschfelder, J. O., Curtiss, C. F. and Bird, R. B. (1954). *Molecular Theory of Gases and Liquids.* Wiley, New York and London.
Johnson, M. F. L. and Stewart, W. E. (1965). *J. Catal.* **4**, 298.
Maxted, E. B. (1945). *Trans. Faraday Soc.* **41**, 406.
Maxted, E. B. (1951). *Adv. Catal.* **3**, 129.
Pease, R. N. and Stewart, L. J. (1925). *J. Amer. Chem. Soc.* **47**, 1235.
Pollard, W. G. and Present, R. D. (1948). *Phys. Rev.* **73**, 762.
Rajadhyaksha, R. A. and Doraiswamy, L. K. (1976). *Catal. Rev. Sci. Eng.* 209.
Ranz, W. E. and Marshall, W. R. (1952). *Chem. Eng. Progr.* **48**, 141.
Rowe, P. N. and Claxton, K. T. (1965). *Trans. Inst. Chem. Eng.* **43**, T231.
Satterfield, C. N. (1970). *Mass Transfer in Heterogeneous Catalysis.* M. I. T. Press, Cambridge, Mass., and London.
Scott, D. S. and Dullein, F. A. L. (1962). *A. I. Ch. E. J.* **8**, 119.
Sharma, C. S., Harriott, P. and Hughes, R. (1975). *Chem. Eng. J.* **10**, 73.
Soomro, M. A. and Hughes, R. (1979). *Can. J. Chem. Eng.* **57**, 24.
Szepe, S. and Levenspiel, O. (1970). *Proc. 4th Europ. Symp. Chem. React. Engng.*, p. 265. Pergamon Press, Oxford.
Thiele, E. W. (1939). *Ind. Eng. Chem.* **31**, 916.
Voorhies, A. (1954). *Ind. Eng. Chem.* **37**, 318.
Wakao, N. and Smith, J. M. (1962). *Chem. Eng. Sci.* **17**, 825.
Weisz, P. B. and Hicks, J. B. (1962). *Chem. Eng. Sci.* **17**, 265.
Weng, H.-S., Eigenberger, G. and Butt, J. B. (1975). *Chem. Eng. Sci.* **30**, 1341.
Wheeler, A. (1951). *Adv. Catal.* **3**, 250.
Wheeler, A. (1955). In *Catalysis*, Vol. 2 (ed. P. H. Emmett), p. 105. Reinhold, New York.
Zeldovitch, J. B. (1939). *Acta Physiochim. (U.S.S.R.)* **10**, 583.

4
Catalyst Deactivation by Sintering

4.1 Introduction

As already indicated in Chapter 2, the sintering behaviour of a catalyst depends on the type of catalyst used. Basically, from the sintering aspect alone, catalysts may be divided into two groups. The first of these may be described as homogeneous in that it is composed of refractory materials (usually an oxide or mixed oxide) and, although in the case of a mixed oxide there will be inhomogeneity on a microscopic scale, the catalyst may be regarded as being homogeneous overall. When such a material is operated at high temperatures over a long period of time, loss of area will occur due to a coalescing of the individual particles or grains making up the compact as described in Fig. 2.2.

The second group of catalysts affected by sintering comprises the supported metal type. In this type of catalyst a metal is present in highly dispersed form on a refractory support. Catalysts of this type may be bifunctional in application, a good example being the platinum reforming catalysts used in the manufacture of gasoline from naphtha fractions. In this particular instance, cracking activity occurs on the alumina support, while hydrogenation and dehydrogenation activity resides in the metal crystallites. For this type of catalyst the active metal crystallites can suffer a reduction in area by a growth of the individual crystallites without any corresponding decrease in the area of the support material. Supported nickel catalysts are another important class of catalysts in this group. Because of the considerable industrial importance of the supported metal catalysts, a great deal of effort has been expended in studies of their sintering behaviour. It will be seen later that there is considerable controversy over the mechanism by which sintering of the crystallites occurs, and it is probable that no theory accounts completely for the phenomena observed in this type of sintering.

In some cases sintering both of support and of the crystallites may occur. This is well illustrated by the work of Williams *et al.* (1972) in an investigation of the sintering of nickel–alumina catalysts. Both total area and nickel area were observed to decrease at temperatures from 400°C to 800°C when treated in steam/hydrogen atmospheres.

Frequently catalyst sintering may occur when the catalyst is regenerated by heating in a weak oxidizing atmosphere to remove deposited coke or carbonaceous material deposited on the catalyst. Usually a mixture containing a few percent of oxygen is employed and the deposit is oxidized to carbon dioxide. Because of the exothermicity of this reaction, great care must be taken during the regeneration operation to control the reaction to prevent too great a temperature rise within the catalyst with consequent sintering. Examples of catalysts that require regeneration include silica–alumina bead catalysts used in processing hydrocarbon feedstocks, the platinum/alumina reforming catalyst referred to above, and cobalt and nickel molybdate catalysts used in hydrodesulphurization reactors. The latter may require more frequent regeneration with increasingly heavier feedstocks being processed. For a cobalt molybdate catalyst the temperature of the combustion front passing through the bed during regeneration should not exceed 550°C if sintering and loss of molybdate from the catalyst are to be avoided. The whole question of catalyst regeneration will be dealt with in detail in Chapter 9; for the present the important point is that regeneration can give rise to sintering.

4.2 Structural features of catalysts relevant to sintering

From a structural viewpoint, loss of area in catalyst compacts is caused by crystal growth through smaller crystals growing into larger crystals with reduction in surface free energy. Metals and metal oxides sinter readily if they are present in the system as very small crystals (<50 nm). A generalization of this concept can be obtained from the data in Table 4.1 where the minimum crystal size that could exist in a sintered compact after appropriate heat treatment is given as a function of temperature (Catalyst Handbook, 1970). Two cases are considered, i.e. a material with a melting point of 1200°C, typical of the supported metals used in catalysis, and a material with a melting point of 2000°C, representative of the support material. Both materials are assumed to have a sintering time of 6 months at the appropriate temperature. Thus a metal of melting point 1200°C would have a minimum

Table 4.1 Temperature (°C) required for crystallites to grow to a given particle size in 6 months.

M.p (°C)	0.01 μm	0.1 μm	1 μm	10 μm	100 μm
1200	125	325	450	530	660
2000	570	730	1020	1270	1680

crystallite size of 0.1 µm if sintered at 300°C for 6 months, whereas alumina (melting point 2032°C) could be held at 500°C for the same period with the crystals not growing beyond 0.01 µm. Because of this, catalysts in which the active constituent is a metal having a relatively low melting point will normally also contain much more refractory material which acts as a "spacer" to impede contact between the more readily sinterable metal constituents. Most spacer materials such as alumina, silica, and chromia can readily be manufactured in a high area form, so these also act as a support material in addition to their role as "spacers".

4.3 Experimental data

4.3.1 Experimental methods

Thermal treatment of a supported metal catalyst results in changes of metal surface area and average metal crystallite size. Some authors (e.g. Wanke and Flynn, 1975) consider sintering as a loss of dispersion of the metal on the support surface. Dispersion is then defined as the ratio of metal surface atoms to total metal atoms. Various methods have been used to measure the loss of metal area; some of the more important are given below.

(a) *Selective gas adsorption*

This is the most obvious and most widely used technique for characterizing the metal area of a supported metal catalyst. The method employs chemisorption of a gas, typically hydrogen, carbon monoxide, or oxygen, although measurements have been made with other gases or vapours. An essential requirement is that the adsorption stoichiometry (i.e. the number of adsorbate molecules adsorbed per surface metal atom) be known. Also required is the number of metal atoms per unit area of surface. The latter will depend on the relevant crystal planes exposed to the gas but, for a polycrystalline material such as dispersed metal crystallites on an oxide support, equal proportions of the low index planes may be assumed (Anderson, 1975). With regard to adsorption stoichiometry there is unfortunately some evidence (Wilson and Hall, 1970) that this may vary with particle size. Nevertheless this method has been widely used because of its simplicity, and if proper precautions are taken there is no reason to doubt that good representations of the relative dispersions of supported metal catalysts can be measured by this technique.

(b) *X-ray diffraction*

There are two main methods for the estimation of the mean particle size: (i) X-ray diffraction line broadening, which relies on information obtained from the peak shape from diffraction lines and is therefore specific for a particular component; (ii) low angle scattering of X-rays, whereby in principle all the particles contribute but the scattering power is dependent on the chemical nature of a component, so a degree of specificity can be achieved.

An X-ray diffraction line broadens when the crystallite size falls below about 100 nm. The X-ray broadening technique is particularly applicable to metal crystallites in the size range 3.0–50 nm. Therefore, if the metal particles are within this range their size may be measured. A detailed examination of the applications of X-ray line broadening has been given by Dorling (1970).

Low angle X-ray scattering has been used for a considerable time to determine particle sizes in the range 1.0–100 nm. Usually the method is applied to radiation scattered within less than 5° of the primary beam. One problem in the application of this method is that low angle scattering from the pores of the support has to be eliminated. Heinemann and coworkers (Whyte et al., 1972) overcame this problem by impregnating the catalyst specimen with a liquid having the same electron density as the support, thus leaving the metal particles as the main scattering centres. Other techniques have been developed (Somoraj, 1968; Renouprez et al., 1974) to minimize this pore scattering effect by compressing or masking of the pores.

(c) *Electron microscopy*

This is the only direct method of determining particle size. Either bright field imaging or dark field imaging transmission electron microscopy may be employed. There are problems due to the interference from the support, but the main problem lies in ensuring that the particle sizes are representative of the catalyst. It is therefore necessary that several samples be examined and a large number of particles measured.

Scanning electron microscopy (SEM) may also be used but, since the limit of resolution is about 15 nm for normal commercial instruments, the method is only really useful for particles larger than about 100 nm.

4.3.2 Experimental results

Most experimental results have concentrated on noble metal (especially platinum) catalysts on an oxide support. The results discussed below necessarily follow this same pattern.

Experimental data have traditionally been fitted to a power law relation. If metal surface area is the variable, the relation may be expressed as:

$$-\frac{dS}{dt} = kS^n \tag{4.1}$$

which, for $n \neq 1$, integrates to:

$$\frac{1}{S^{n-1}} - \frac{1}{S_o^{n-1}} = kt \qquad (4.2)$$

where S_o is the initial metal area.

If the mean particle diameter d or radius $\bar{r}$ is employed to measure the degree of sintering, then:

$$\frac{\mathrm{d}(\bar{r})}{\mathrm{d}t} = k(\bar{r})^{-m} \qquad (4.3)$$

or

$$(\bar{r})^{m+1} - (\bar{r}_o)^{m+1} = kt \qquad (4.4)$$

where $\bar{r}$ and S are connected by the relation:

$$\bar{r} = \frac{3V}{S} \qquad (4.5)$$

If spherical particles are assumed, the exponents in the particle radius and area equations are related by:

$$m = n - 2 \qquad (4.6)$$

When data are obtained in terms of increased particle radius, equation (4.4) may be written more conveniently in logarithmic form:

$$(m + 1)\log\left(\frac{\bar{r}}{\bar{r}_o}\right) = \log\left(1 + \frac{kt}{(\bar{r}_o)^{m+1}}\right) \qquad (4.7)$$

If $kt/(\bar{r}_o)^{m+1}$ is much greater than unity, equation (4.7) may be written:

$$(m + 1)\log\left(\frac{\bar{r}}{\bar{r}_o}\right) \approx \frac{k}{(\bar{r}_o)^{m+1}} + \log t \qquad (4.8)$$

Thus a plot of $(\bar{r}/\bar{r}_o)$ against t on logarithmic paper should give a straight line of slope $1/(m + 1)$, which is equivalent to $1/(n - 1)$ if the approximation used in obtaining equation (4.8) holds.

Attempts have been made to prove or disprove a particular sintering mechanism by fitting data to a single value of the exponents n or m. However, there are dangers in such a course, since average properties of the metal particles are assumed and it could be that in actual sintering a sequence of processes occurs with the possibility of a changing rate exponent.

Nevertheless this approach has proved useful in providing a basis on which theories of the mechanism of sintering may be compared; if the values of the exponents n or m are not large, such a procedure may be useful.

Most data available in the literature have been obtained in laboratory reactors with total sintering times of a few weeks and more typically of about 4–5 days only. Thus the data may not be wholly representative of plant conditions and would probably not be complicated by the phenomenon of redispersion which sometimes occurs (see below).

Table 4.2 summarizes the experimental information available on the power law orders. As indicated above, for many investigations a range of values for n was obtained. The most obvious conclusion from Table 4.2 is that a very large variation in the values of n is obtained and also that n can have very high values (up to 16). Results are given for the various atmospheres employed in these sintering experiments and it can be seen that there is a general trend for the order, n, to be larger in reducing atmospheres than in oxidizing atmospheres. The orders in nitrogen are approximately the same as those in air. It has also been generally observed that higher values of n occur when the metal is initially in a very fine state of dispersion on the surface.

Another parameter important in mechanistic interpretations of sintering is the apparent activation energy of the process. Data have been obtained from

Table 4.2 Power law orders (n) from sintering rates.

Catalyst	Conditions	n	Reference
5% Pt/Al_2O_3	$H_2/600°C$	12–16	Somoraj (1968)
	700	14–16	
1.1% $Pt/\eta\text{-}Al_2O_3$	$H_2/500$	6–7	Gruber (1962a)
0.7% $Pt/\eta\text{-}Al_2O_3$	$H_2/500$	15	Gruber (1962b)
0.6% $Pt/\eta\text{-}Al_2O_3$	$H_2/500$	11	
0.4% $Pt/\eta\text{-}Al_2O_3$	$H_2/482$	9	Hughes et al. (1962)
	$H_2/538$	8	
0.774% $Pt/\gamma\text{-}Al_2O_3$	$N_2/564$	2–3	Hermann et al. (1961)
	$N_2/594$	2	
	$N_2/625$	2–5	
0.375% $Pt/\gamma\text{-}Al_2O_3$	$N_2/564$	2	Hermann et al. (1961)
	$N_2/594$	2	
	$N_2/625$	2–5	
5.0% Pt/carbon	$N_2/600$	5–12	Bett et al. (1974)
	$N_2/700$	5–14	
	$N_2/800$	5–8	
12% Pt/carbon	$N_2/600$	4–7	Bett et al. (1974)
$Pt/\gamma\text{-}Al_2O_3$	2% O_2 in $N_2/700$	12–13	Wynblatt (1973)
$Pt/\gamma\text{-}Al_2O_3$	Air/700	5	
$Pt/\gamma\text{-}Al_2O_3$	Air/900	5	Huang (1973)
5% Pt/Al_2O_3	Air/600	9–10	Somoraj (1968)
	Air/700	8–14	
0.6% $Pt/\gamma\text{-}Al_2O_3$	Air/780	2	Maat and Moscou (1965)

4 Catalyst Deactivation by Sintering

two types of experiment: (a) variable time, variable temperature studies; (b) constant time, variable temperature studies. Estimates of the activation energy of sintering for these two different types of experiment estimated by Wanke and Flynn (1975) are given in Tables 4.3 and 4.4 respectively. For constant time and variable temperature a value for the exponent n must be assumed. The activation energies are presented in Table 4.4 for various values of n. It should be noted that higher values of n always result in large values of the apparent activation energy.

The results in Tables 4.3 and 4.4 show that, in general, apparent activation energies for sintering are higher in oxidizing or inert atmospheres than in reducing atmospheres. The only notable exceptions are the 1 % Pt/SiO_2 and the 5 % Rh/SiO_2 quoted in Table 4.4. However, it appears from an examination of the original papers that these catalysts were not completely reduced prior to sintering. Hence, the low activation energies obtained may be due to migration of the incompletely reduced metal salt, causing large metal salt crystals which on subsequent reduction result in low metal dispersions.

It may be concluded that activation energies for sintering are generally high. Also, the range of power law exponents obtained emphasizes the difficulty of fitting mechanistic expressions. The large values of n, in particular, are difficult to explain on conventional deactivation theory and have caused considerable controversy among the advocates of different sintering models.

Factors affecting the rate of sintering

The most important factors affecting the rate of sintering are temperature and the atmosphere in which sintering is conducted. In contrast, the nature of the support and degree of metal loading appear to be of secondary importance only.

The rate of sintering, as expected, always increases with increasing

Table 4.3 Apparent activation energies (kJ mol^{-1}) for sintering; variable time, variable temperature studies.

Catalyst	Conditions	E_a	Reference
0.4% Pt/Al$_2$O$_3$	H$_2$/482–538°C	62.7	Hughes et al. (1962)
5% Pt/Al$_2$O$_3$	H$_2$/600–700	83.6	Somoraj (1968)
5% Pt/C	N$_2$/600–800	175.6	Bett et al. (1974)
0.774% Pt/γ-Al$_2$O$_3$	N$_2$/564–625	292.6	Hermann et al. (1961)
0.375% Pt/γ-Al$_2$O$_3$	N$_2$/564–625	355.3	
5% Pt/Al$_2$O$_3$	Air/600–700	217.4	Somoraj (1968)

Table 4.4 Apparent activation energies (kJ mol^{-1}); constant time, variable temperature experiments.

Catalyst	Conditions	Power law order				Reference
		2	6	10	14	
3.7% Pt/η-Al$_2$O$_3$	Vacuum/600–800°C				209	Renouprez et al. (1974)
1.6% Pt/SiO$_2$	H$_2$/700–800		125	167		Benesi et al. (1968)
1% Pt/SiO$_2$	Air/400–700	54	146	209		Sagert and Ponteau (1971)
1% Pt/SiO$_2$	Air/350–600	25				
0.5% Pt/Al$_2$O$_3$	Air/600–800	146	188			Joworska-Galas and Wrzyszes (1966)
0.4% Pt/Al$_2$O$_3$	Air/650–700	251	710			
5% Rh/SiO$_2$	Air/538–800	42	146			Yates and Sinfelt (1967)
5% Pd/Al$_2$O$_3$	H$_2$/600–800	42	167	314		Aben (1968)
2% Pd/SiO$_2$	H$_2$/600–900	71	125			
10% Ni/Al$_2$O$_3$–SiO$_2$	H$_2$/500–700	25	85	167	251	Carter et al. (1968)

temperature. Also, the rate of sintering is larger in oxygen containing atmospheres than in those containing hydrogen, and the latter is comparable to the rate in nitrogen atmospheres (Somoraj, 1968). Sintering in oxygen containing atmospheres decreases with decrease in the oxygen partial pressure (Wynblatt and Gjostein, 1973).

One phenomenon of great practical importance, for which no satisfactory explanation has been given, is that of redispersion of a catalyst that has already suffered loss of active area by sintering. There are a number of processes given in the patent literature whereby this desirable end may be achieved. However, in all cases the atmosphere must contain some oxygen. One patent (Netherlands, 1968) advocates regeneration of a supported platinum catalyst deactivated through sintering by heating at 370–550°C in an inert gas stream containing 0.5–2% oxygen. In this particular case the regenerated catalyst had a greater activity than the original fresh catalyst.

It is often important to be able to assess the relative resistance to sintering of the various metals used in supported form in catalysts. This is difficult to determine unequivocally because of the different treatments reported in the various experimental studies, but it seems that in non-oxidizing atmospheres resistance towards sintering increases with increasing melting point of the metal, as might be expected. The order of stability in non-oxidizing atmospheres therefore increases in the sequence $Ni < Pd < Pt < Rh$. In oxidizing atmospheres no conclusions can be drawn from current data.

4.4 Mechanisms of sintering for supported metal catalysts

In general, growth of a metal crystallite on an oxide support can occur by two distinct mechanisms: (i) particle migration and coalescence; (ii) transfer of metal atoms individually from one particle and their deposition on another. These processes have been the subject of much discussion and have been reviewed in an excellent paper by Wynblatt and Gjostein (1975) who have also proposed a theory of "inhibited" growth of particles which takes account of the faceted nature of the crystallites on the support surface. The model based on particle migration and coalescence was proposed by Ruckenstein and Pulvermacher (1973a,b), while that of interparticle transport was developed by Flynn and Wanke (1974a,b). Both are described in detail later in this chapter, but a brief outline of the reasoning leading to the development of each model is given below.

4.4.1 Particle migration and coalescence

The discussion that follows does not include metallic clusters of a few atoms only, which migrate by different mechanisms, but is limited to the more usual crystallite particles typical of a supported platinum catalyst. The particle surface diffusivity, D_p, can be related to the diffusivity, D_a, of the metal atoms on the particle surface. Two models have been proposed to obtain this relation. The first assumes spherical particles with no surface energy anisotropy and is based on the analogy of surface diffusion controlled migration of voids in solids proposed by Gruber (1967). This gives:

$$D_p = 4.816 D_a (a/d)^4 \qquad (4.9)$$

where a is the atom diameter and d that of the particle.

Alternatively, if a faceted particle is assumed, so that the process is controlled by the nucleation of new monatomic layers on the facets, application of nucleation theory (Wynblatt and Gjostein, 1975) gives:

$$D_p = (D_a d/2l) \exp[-(\pi g \gamma_e / kT) d] \qquad (4.10)$$

where l is the jump distance, g is the ratio of facet diameter to particle radius, and γ_e is the edge energy of a two-dimensional array of metal atoms.

The relation between D_p and particle diameter may be determined from both (4.9) and (4.10) by taking appropriate values for parameters. Thus for equation (4.9), and taking data for platinum at 600°C, $a = 0.277$ nm and $D_a = 5.0 \times 10^{-14}$ m² s^{-1}, D_p is found to decrease with particle diameter from about 10^{-15} m² s^{-1} at 1 nm to about 10^{-18} m² s^{-1} at 10 nm. Similar calculations with equation (4.10), using $\gamma_e = 2 \times 10^{-11}$ J m^{-1}, $g = 0.5$, and assuming $l = 0.277$ nm, give a somewhat steeper dependence on particle size, D_p decreasing from about 10^{-15} m² s^{-1} at 2 nm to 10^{-23} m² s^{-1} at 10 nm. However, approximate agreement between the two models is obtained for particle diameters less than 6 nm. For estimation purposes D_p may be assumed to be 10^{-15}–10^{-16} m² s^{-1} for platinum particles of 2 nm diameter, while for 6 nm diameter particles D_p is about 10^{-18}–10^{-19} m² s^{-1}.

4.4.2 Interparticle transport

Two different growth laws have been proposed by Wynblatt and Gjostein (1975) for interparticle transport, depending on whether transport of atoms across the particle edge or atom diffusion across the support is rate controlling. Both have the form:

$$(\bar{d})^r \propto t \qquad (4.11)$$

where r is equal to 4 for diffusion control and 3 for edge control.

The concentration of diffusing atoms is controlled by the energetics of the transport processes involved, i.e. the heat of atomization (ΔH_m) minus the atom–support interaction (ΔH_s). If ΔH_s is only equal to the physical adsorption (van der Waals) energy, the atom concentration will be negligible at normal sintering temperatures and would only become important if ΔH_s is at least half of ΔH_m. This requirement is unlikely to be achieved on clean oxide supports but under oxidizing conditions ΔH_s is increased with noble metals because of the formation of mobile molecules of metal oxides, and under these conditions interparticle transport can become feasible either by surface or vapour transport. For vapour transport equation (4.11) still holds but the value of r is now 2.

Abbreviated forms of the particle migration model and the atom migration model are given in the following two sections. These are then critically discussed.

4.4.3 The particle migration model

This model was proposed by Ruckenstein and Pulvermacher (1973a,b) as a development of earlier work by Smoluchovski (1917) and Chandrasekhar (1943) on coagulation theory. The concept of particle migration was based on the reasoning that the platinum crystallites are at most 1 nm in size and interactions between the platinum atoms and the support are weaker than the interactions between platinum atoms themselves. It is therefore assumed that, for temperatures higher than the Tamman temperature (0.4 times the melting point expressed as K), the crystallites are in a quasi-liquid state. Hence they may migrate as crystallites on the surface of the support.

There are two different types of mechanism arising from such a process, depending on the nature of the coalescence between two particles when these approach one another. If the interaction between the particles is assumed to be so strong that two particles in contact form a single unit during a time that is short compared with the time in which migration takes place, the process of coalescence will be diffusion controlled. If the merging process is slow, however, when compared with the diffusion process, "sintering control" will be rate determining. The following development is taken from Ruckenstein and Pulvermacher (1973a).

Kinetic development

Let C_k be the number of crystallites per unit area of support, each composed of k platinum atoms. The concentration C_k will increase by collision of particles consisting of i and $(k - i)$ atoms and will decrease by collision of particles containing k atoms with any other particles.

The rate of collision between particles comprising i and j units is:

$$b_{ij} = K_{ij} C_i C_j \qquad (4.12)$$

where K_{ij} are the rate constants dependent on the mobility of the crystallites and on the interaction between the particles.

The rate of change of C_k is thus:

$$\frac{dC_k}{dt} = \frac{1}{2} \sum_{i+j=k} K_{ij} C_i C_j - C_k \sum_{i=1}^{\infty} K_{ij} C_i \qquad (4.13)$$

The collision rate of the particles may be estimated as follows. If the centre of the contact surface between a particle composed of j units and the support is taken as the origin of a cylindrical coordinate system, the migration of particles consisting of k units upon the support surface is due to thermal motion of the atoms. It therefore has a random character and may be represented by a diffusional model; therefore:

$$\frac{\partial C_k}{\partial \theta} = D_{kj} \left[\frac{\partial^2 C_k}{\partial r^2} + \frac{1}{r} \frac{\partial C_k}{\partial r} \right] \qquad (4.14)$$

where D_{kj} is the diffusivity of particle k with respect to that of particle j. In equation (4.14) θ denotes a small time scale, whereas t in (4.13) is a longer time scale. Equation (4.14) is used to obtain the collision rate between particles during a time interval θ_o that is small compared with the process time t in which appreciable sintering occurs by particle size redistribution.

The particle consisting of j units is also subject to movements that are independent of the movements of the k particles. The diffusivity D_{kj} is thus given by:

$$D_{kj} = D_k + D_j \qquad (4.15)$$

The following initial boundary conditions are relevant:

$$\text{At } \theta = 0, \quad C_k = C_{k0}, \quad r > R_{kj} \qquad (4.16a)$$

$$\text{For } \theta > 0, \quad \alpha_{kj} C_k = D_{kj} \frac{\partial C_k}{\partial r}, \quad r = R_{kj} \qquad (4.16b)$$

where α_{kj} is the reaction rate constant for the merging process, and R_{kj} is the radius of interaction of the two colliding particles, which may be assumed to be:

$$R_{kj} = r_k + r_j \qquad (4.17)$$

For very high coalescence rates the process is diffusion controlled and the concentration C_k for $R_{kj} = r_j + r_k$ is zero. Conversely, when reaction rates for coalescence are sufficiently low compared with the diffusivity, the process is sintering controlled.

The rate of collision, Φ_{kj}, is equal to the product of the diffusion flux and the perimeter, $2\pi R_{kj}$:

$$\Phi_{kj} = \left[2\pi R_{kj} D_{kj} \left(\frac{\partial C_k}{\partial r} \right)_{R_{kj}} \right] \qquad (4.18)$$

4 Catalyst Deactivation by Sintering

Solutions to equations (4.18) and (4.14) were obtained by Ruckenstein and Pulvermacher (1973a). The full expressions are given in the original paper; here only the simplified forms are quoted. For diffusion control K_{ij} is given by the expression:

$$K_{ij} = \frac{8D_{ij}}{\pi} \int_0^\infty e^{-D_{ij}u^2\theta} \frac{du}{[J_o^2(uR_{ij}) + Y_o^2(uR_{ij})]u} \quad (4.19)$$

while for sintering control:

$$K_{ij} = 2\pi R_{ij}\alpha_{ij} \quad (4.20)$$

For sintering control the rate constants are independent of the small scale time θ because the latter affects only the rate of diffusion of particles towards a selected particle. In this situation diffusion is not rate determining. For diffusion control the rate constants depend on the small scale time θ, but in many cases this dependence becomes very weak after a very short time interval and

$$K_{ij} = \frac{4\pi D_{ij}}{\ln 4T} \quad (4.21)$$

becomes a good approximation for equation (4.19). Here T is the dimensionless time.

The kinetic equation (4.13) may be solved for the limiting cases of diffusion control or kinetic (sintering) control. Equation (4.21) is used to evaluate K_{ij} for diffusion control because calculation of the dimensionless time T shows that this is very large after a short time θ. If D_{ij} is considered to be analogous to a surface diffusion coefficient, it should have a magnitude between 10^{-8} and 10^{-12} m² s⁻¹ (Satterfield, 1970). The collision radius is of the order of the particle sizes in the system, i.e. 10^{-8}–10^{-9} m. Hence, T is about $(10^2$–$10^{10}) \times \theta$ and is large even for small values of θ. Therefore, $\ln 4T \gg 1$ and equation (4.21) is a good approximation for (4.19). Hence, for the diffusion controlled case:

$$\frac{dC_k}{dt} = \frac{1}{2} \sum_{i+j=k} M_{ij} 4\pi D_{ij} C_i C_j - C_k \sum_{i=1}^\infty M_{ki} 4\pi D_{ik} C_i \quad (4.22)$$

where

$$M_{ij} = \frac{1}{\ln(4D_{ij}\theta/R_{ij}^2)} \quad (4.23)$$

and for the case of sintering control:

$$\frac{dC_k}{dt} = \frac{1}{2} \sum_{i+j=k} 2\pi R_{ij}\alpha_{ij} C_i C_j - C_k \sum_{i=1}^\infty 2\pi R_{ik}\alpha_{ik} C_i \quad (4.24)$$

Deactivation of Catalysts

The diffusion coefficient and the reaction rate constants may depend on the size of the particles. It is expected that for diffusion control the rate constants K_{ij} decrease with increasing particle size, and therefore the smallest particles in the system are the fastest to disappear. For sintering control the rate constants K_{ij} will increase with increasing particle size, and therefore the smallest particles in the system disappear most slowly. The variation of K_{ij} with particle size may be expressed by the general relation:

$$K_{ij} = C[r_i^{3m} + r_j^{3m}] \tag{4.25}$$

where C is independent of particle size.

The exposed surface area of the metal crystallites is given by:

$$S = \sum_{k=1}^{\infty} S_k C_k \tag{4.26}$$

where S_k is the exposed surface area of the particle containing k units. Assuming that the shape of the particles does not change during the growth process, the ratio of volume to exposed area is proportional throughout the sintering process to the first power of some characteristic length of the particle; therefore:

$$S_k = S_1 k^{2/3} \tag{4.27}$$

Table 4.5 Time evolution of exposed metal surface area in the two limiting cases.

Diffusion control (assumed size dependence of D_{ij})	Rate constants	Rate equation for exposed surface area
$D_{ij} = C_1\left(\dfrac{1}{r_i^2} + \dfrac{1}{r_j^2}\right)$	$K_{ij} = C'_1\left(\dfrac{1}{r_i^2} + \dfrac{1}{r_j^2}\right)$	$\dfrac{dS}{dt} = -C''_1 S^6$
$D_{ij} = C_2\left(\dfrac{1}{r_i} + \dfrac{1}{r_j}\right)$	$K_{ij} = C'_2\left(\dfrac{1}{r_i} + \dfrac{1}{r_j}\right)$	$\dfrac{dS}{dt} = -C''_2 S^5$
$D_{ij} = C_3$	$K_{ij} = C'_3$	$\dfrac{dS}{dt} = -C''_3 S^4$

Kinetic control (assumed size dependence of α_{ij})	Rate constants	Rate equation for exposed surface area
$\alpha_{ij} = C_4$	$K_{ij} = C'_4(r_i + r_j)$	$\dfrac{dS}{dt} = -C''_4 S^3$
$\alpha_{ij} = C_5 \dfrac{r_i^2 + r_j^2}{r_i + r_j}$	$K_{ij} = C'_5(r_i^2 + r_j^2)$	$\dfrac{dS}{dt} = -C''_5 S^2$

where S_1 is the exposed surface area of a particle containing one unit. The rate of change of total crystallite surface area with time is then:

$$\frac{dS}{dt} = S_1 \sum_{k=1}^{\infty} k^{2/3} \frac{dC_k}{dt} \qquad (4.28)$$

The exposed surface area given by equation (4.28) was computed numerically by Ruckenstein and Pulvermacher assuming various size dependences of D_{ij} and α_{ij} for the cases of diffusion control and sintering control respectively. The various assumed dependences and the derived rate constants and equations for exposed surface area are shown in Table 4.5.

It will be observed that, for the sintering control case, exponents of 2 and 3 are obtained for the fixed surface area rate expressions whereas in the case of diffusion control the exponent is 4 or larger.

4.4.4 The atomic migration model

Wanke and Flynn (1975) have criticized the particle migration model on several grounds. In particular they suggest, from the theoretical evidence of Wynblatt and Gjostein (1975), that sintering control cannot be rate controlling at temperatures above 500°C, i.e. sintering control is not generally realizable under normal sintering conditions. They also suggest that during the initial stages of sintering some particle migration may occur, but this is unlikely to be the dominant mechanism for larger particles. They cite evidence (Flynn and Wanke, 1974a) that platinum agglomeration continues even when the size of the metal crystallites exceeds the size of the alumina support particles. They therefore concluded that during the latter stages of sintering an alternative mechanism was necessary, and accordingly produced a model based on atomic migration (sometimes called the interparticle transport model).

Essentially, the process as envisaged by Flynn and Wanke (1974b) comprises three steps: (a) transport of metal atoms from the crystallite to the surface of the support; (b) migration of metal atoms on the support surface; (c) capture of migrating atoms by collision with a metal crystallite or immobilization of the atoms by a decrease in temperature or by encountering an energy sink on the support surface. These three processes are now discussed in turn.

(a) *Escape of metal atoms from crystallites*

The possibility of metal atom migration had been discounted prior to Flynn and Wanke's papers (1974a,b) because of the unfavourable energetics of a process involving removal of metal atoms from a crystallite to the surface. Thus the heat of sublimation of platinum is 564 kJ mol^{-1} while the adsorption of the metal atoms has generally been assumed to be of the van der Waals type and therefore the heat of adsorption should be less than 83 kJ mol^{-1}. Sintering of metal crystallites, such as platinum, generally occurs at temperatures between 500 and 700°C, so an activation energy less than 250–310 kJ mol^{-1} is necessary if sintering is to occur at a significant rate within this temperature range. It is therefore necessary to postulate an increased interaction of the metal atom and the support surface to reduce the activation energy below 310 kJ mol^{-1}. It was observed by Geus (1971) that the presence of defects and/or the presence of cracking products from hydrocarbons would increase the metal–support interaction. From this and other indirect evidence, Flynn and Wanke proposed that sufficient adsorption energy could be available to reduce the activation energy below the upper bound of 310 kJ mol^{-1}.

The reduction of surface energy is the driving force for the migration of metal from larger to smaller crystallites. An approximate relation for the rate of loss of metal atoms from the crystallite to the support as a function of crystallite size may be obtained as follows. The spreading pressure, ψ, is related to the crystallite radius, r, by the Kelvin equation:

$$\psi = \psi_o \exp(\beta/r) \tag{4.29}$$

where β depends on the crystallite shape, the metal–support contact angle, the metal–support and metal–vapour interfacial energies, the metal molar volume, and the temperature. For a specific system at constant temperature, β is approximately constant. For a single crystallite in equilibrium with free atoms on the support surface, the rate at which the crystallite captures atoms is proportional to $r\psi$, and therefore the rate of atom loss is also proportional to $r\psi$. Hence, the ratio of the rates of loss for two crystallites of different size is given by:

$$\frac{dL_1}{dt} \bigg/ \frac{dL_2}{dt} = \frac{r_1 \psi_1}{r_2 \psi_2} \exp\left[\beta\left(\frac{1}{r_1} - \frac{1}{r_2}\right)\right] \tag{4.30}$$

The contact angle for Pt/Al$_2$O$_3$ is 90°, so the crystallites were assumed to be hemispherical. For this case β is given by:

$$\beta = (\gamma_1 + 2\gamma_2)V/RT \tag{4.31}$$

where γ_1 and γ_2 are the metal–support and metal–atmosphere interfacial energies and V is the metal molar volume. Using this relation for selected values of β, the relative rates of metal atom loss were calculated for crystallites from 2 to 50 nm using equation (4.30). On this basis the rate of loss of metal atoms from a crystallite was found to be independent of crystallite size except for crystallites of size 2 nm or less. (This however may be the range of interest for many crystallites.) Therefore the rate of loss of

4 Catalyst Deactivation by Sintering

atoms from the ith particle is:

$$\frac{dL_i}{dt} = A\, e^{-E_i/RT} \tag{4.32}$$

where E_i is the activation energy to move an atom from the crystallite to the surface. This relation does not hold for particles of 30 nm and above.

(b) *Migration of metal atoms over support surfaces*

Evidence from nucleation and film growth studies shows that metal atoms are mobile on support surfaces. The atoms may be considered as a two-dimensional gas whose velocity is given by:

$$v = \left(\frac{\pi k_B T}{2m}\right)^{\frac{1}{2}} \tag{4.33}$$

or if a "jump" mechanism from one surface site to another is considered:

$$v = av\exp(-E_s/RT) \tag{4.34}$$

where E_s, the activation energy for surface diffusion, is usually much less than E_i.

Either equation can be used in Flynn and Wanke's model.

(c) *Capture of atoms by crystallites*

The rate at which a crystallite gains metal atoms, which depends on the concentration of metal atoms on the support surface, the velocity of these atoms, and the effective diameter of the crystallites, D_i, is given by:

$$\frac{dG}{dt} = \alpha\left(\frac{F_s}{N_t S_o}\right)v D_i \tag{4.35}$$

where α is the atom/crystallite sticking probability and F_s is the number of atoms migrating on a support with a total area of $N_t S_o$ (N_t is total metal atoms and S_o the area per atom).

Summary of the model

The net change of the number of atoms in the ith particle is given by:

$$\frac{dN_i}{dt} = \frac{dG_i}{dt} - \frac{dL_i}{dt} \tag{4.36}$$

which may be written in expanded form:

$$\frac{dN_i}{dt} = \alpha v D_i\left(\frac{F_s}{N_t S_o}\right) - A\, e^{-E_i/RT} \tag{4.37}$$

The rate of change of the number of migrating atoms is given by the material balance:

$$\frac{dF_s}{dt} = \sum_{i=1}^{M} \left(\frac{dL_i}{dt} - \frac{dG_i}{dt} \right)$$

$$= -MA\, e^{-E_a/RT} + \alpha v \left(\frac{F_s}{N_t S_o} \right) \sum_{i=1}^{M} D_i \qquad (4.38)$$

where M is the number of crystallites on the support area $N_t S_o$. These last two equations were solved numerically for various particle size distributions by Flynn and Wanke taking various values of the parameters α, A, and E_i.

4.4.5 Comparison of models for sintering

Since a crystallite may be presumed to have many bonds between the metal and support atoms, which necessarily must be broken before crystallite migration can occur, while a free metal atom requires fewer bonds to be broken, it might be thought that crystallite diffusion is slow compared with free atom transport. However, the spacings of metal and support atoms are such that only a fraction of metal atoms on the crystallite base are in an energy well between atoms of the support. In contrast, free metal atoms can almost always be placed without hindrance between support atoms. Hence, although the number of metal–support bonds may be much greater in the case of a large crystallite, the strength of each bond is less than that for a single metal atom. Thus crystallite motion is possible. Also, crystallites having a significant size may be able to overcome obstacles such as valleys on the irregular support surfaces, which would probably trap and immobilize single atoms.

There is also experimental evidence for crystallite migration from electron microscopy measurements (Bassett, 1960, 1964). Copper and silver "islands" on carbon supports were observed to move at high temperatures, and translation of silver "islands" on molybdenum occurred at temperatures as low as 523 K. Motion and coalescence of gold "islands" on amorphous carbon and silicon at temperatures between 500 and 700 K has also been reported by Skofronick, Phillips, and coworkers (1967, 1968). Crystallites of supported metals are 1–10 nm in size, which is approximately the same diameter as the "islands" in the above experiments, so crystallite migration can be considered to be feasible.

Arguments against the particle migration model include the evidence provided by Flynn and Wanke (1974a) that metal crystallites can grow to exceed the size of the support particles and can grow from one particle to another. This is difficult to explain by particle migration, whereas atomic diffusion across the contact points of the support particles would seem to be possible under these conditions. Vapour transport could also account for this phenomenon. Another argument used against the particle migration model is that it does not account quantitatively for values of n up to 15 as found experimentally. For example, Wynblatt and Gjostein report orders of 10–13 for platinum crystallite migration on a plane support surface, whereas on the particle migration model, as originally conceived, an upper limit of n equal to 8 was given by Ruckenstein and Pulvermacher (1973b). More recently Ruckenstein and Dadyburjor (1977) pointed out that this difficulty can be removed if it is assumed that the diffusivity is inversely proportional to higher powers of the particle size; if this is true, higher orders for sintering will be obtained.

A fundamental problem in the atomic migration model is the acquisition of the necessary energy for the atom to be removed from the crystallite to the surface of the support. As already pointed out, the high heat of sublimation of platinum (564 kJ mol^{-1}), coupled with an estimated value of the heat of adsorption for platinum monomer on a substrate of 38 kJ mol^{-1} (McLean and Hondros, 1971), puts limits on the activation energy necessary for sintering to occur at a significant rate. This activation energy for migration of a platinum atom over the support must necessarily lie between 0 and the heat of adsorption. Wynblatt et al. (1975) have shown that taking the upper bound for this activation energy (i.e. equal to the heat of adsorption) and inserting appropriate parameters in the crystallite growth expressions gives a negligibly small increase in size of about 0.1 nm in 10^{18} s. This infinitesimal rate of growth is due to the very large heat of sublimation of platinum and the small heat of adsorption of platinum atoms on the substrate.

However, this situation can change when impurities are present on the surface and, more important, it can be considerably modified under oxidizing conditions. The surface of platinum in an oxygen containing atmosphere will have some coverage of chemisorbed oxygen, probably as a surface oxide. At a sufficiently high temperature, and when equilibrium is achieved, there will be a significant vapour pressure of PtO_2 in the gas phase, given by the equilibrium constant for the reaction:

$$Pt(s) + O_2(g) \rightleftharpoons PtO_2(g) \quad (4.39)$$

Under these conditions the energy required to remove a platinum atom to the gas phase (in the form of a PtO_2 molecule) will be the enthalpy change for (4.39), i.e. about 176 kJ mol^{-1}, which is considerably less than the heat of sublimation of platinum. Estimates of growth of platinum particles, made using this concept of PtO_2 being the transport medium, are reported in detail below. However, the rates obtained are now quite fast and often exceed the experimental values. This led Wynblatt and Gjostein (1975) to propose a theory of inhibited growth by a nucleation mechanism on faceted crystallites.

Table 4.6 Platinum particle migration distances at 600°C; $t = 5$ h, $D_a = 5 \times 10^{-14}$ m^2 s^{-1} at 600°C.

d (nm)	D_p (m^2 s^{-1})	$x_p = 2\sqrt{D_p t}$ (nm)
5	1.5×10^{-18}	330
20	5.9×10^{-21}	20
50	1.5×10^{-22}	3.3
500	1.5×10^{-26}	0.033

The importance of particle migration may be assessed in a semi-quantitative way by comparing calculated average crystallite migration distances with average interparticle spacings in a typical supported platinum catalyst. For example, a 1% platinum catalyst with platinum particles of average diameter 6 nm on a support of area 200 m^2 g^{-1} would possess an average interparticle spacing of 200 nm. If the average particle migration distance was then much less than the average interparticle spacing, particle migration could be neglected as a growth process. This root mean square platinum particle migration distance may be estimated from D_p (the particle diffusivity) through the conventional relation for two-dimensional diffusion:

$$x_p = 2\sqrt{D_p t} \qquad (4.40)$$

The value of D_p decreases rapidly with increasing particle size:

$$D_p = 4.816 D_a (a/d)^4 \qquad (4.41)$$

Estimates of D_p and the corresponding average migration distances, x_p, are given in Table 4.6. Taking a value for $d = 6$ nm of $D_p = 3 \times 10^{-19}$ m^2 s^{-1} at 600°C gives a value for x_p of 147 nm. Therefore it appears that a diameter of 10 nm is the upper bound on size beyond which particle migration makes no significant contribution to particle growth.

The above arguments refer to spherical particles. When the platinum particles are faceted (which seems to occur under oxidizing conditions) similar calculations by Wynblatt and Gjostein (1975) give an upper limit to the particle size for which particle migration can occur of 9 nm. Thus both analyses on spherical and faceted particles suggest strongly that particle migration will occur at small particle size.

4.4.6 Experimental work of Wynblatt and Gjostein

Wynblatt and Gjostein (1973) and Wynblatt et al. (1975) have measured sintering in model systems consisting of both flat and microporous supports. Results obtained initially on their films of alumina containing platinum

4 Catalyst Deactivation by Sintering

particles gave values for the order of sintering (n) equal to 12. It was also demonstrated (Wynblatt and Gjostein, 1973) that when $\log \bar{r}$ was plotted against $\log t$ the plot was linear when the atmosphere contained 2% oxygen in nitrogen, but when air was used upward curvature was exhibited, showing the enhancement of sintering caused by an appreciable oxygen partial pressure.

The atomic migration model was compared with experimental results on both flat and microporous supports in the work published in 1975 and Fig. 4.1 shows the results obtained. It will be noted that for both supports the theory predicted sintering rates well above those obtained experimentally. Similar comparisons between theory and experiment for the same oxidizing environment ($pO_2 = 0.2$ atm) were made using the particle migration model and the results are shown in Fig. 4.2. It can be observed that the values from theory are slightly lower than observed values of sintering for the plane support surfaces when $pO_2 = 0.02$ atm, and appreciably lower for $pO_2 = 0.2$ atm. Nevertheless the agreement is better than for the atomic migration model.

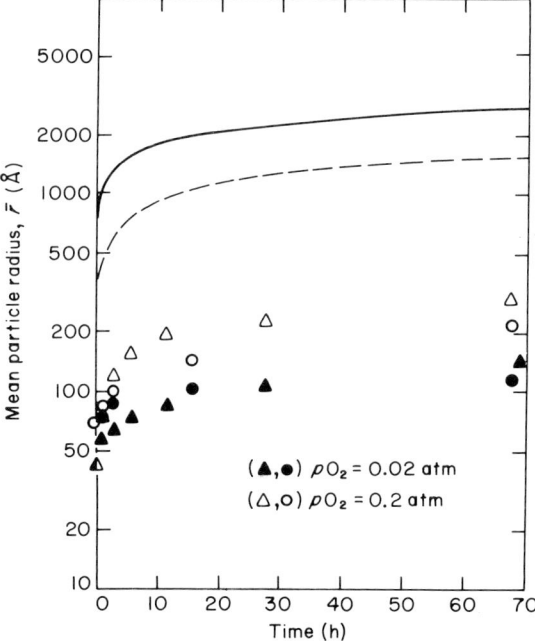

Fig. 4.1 Comparison of all data with estimates (curves) of non-inhibited growth by diffusion over the substrate. Note logarithmic scale of $\bar{r}$ axis. (△, ▲) Microporous support (solid surve). (○, ●) Flat support (dashed curve). (Wynblatt et al., 1975)

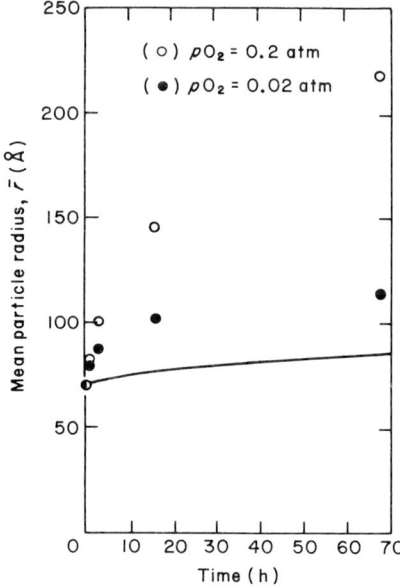

Fig. 4.2 Comparison of data from flat-support experiments with estimate (curve) of non-inhibited growth by particle migration, collision, and coalescence. (Wynblatt et al., 1975)

The lower results obtained experimentally than were predicted by the atomic migration model led Wynblatt and Gjostein to postulate an inhibited growth model. It is well established from crystal growth theory that the development of singular interfaces can inhibit growth rates because of the necessity to "nucleate" fresh surfaces for continued growth. In sintering, facets were found to develop on the crystallites as sintering proceeds. The inhibition caused by the faceting process on the crystallites does not influence to any great extent the particle migration of small crystallites which develop during the early stages of sintering, but it does have a considerable influence on the atomic migration processes. Faceting causes a retardation in the latter process and brings the model predictions into general agreement with those obtained experimentally. Wynblatt and Gjostein have described this type of process as "inhibited growth".

Wynblatt and Gjostein have also considered this process as a mechanism for sintering of supported and metal catalysts. Such a process would be controlled by the interface mechanism since diffusion through the gas phase is rapid. Assuming an oxidizing environment with PtO_2 molecules involved in the adsorption and desorption processes at the platinum particle

4 Catalyst Deactivation by Sintering

interfaces, rates of sintering may be computed and these have been compared with experimental results. Much lower growth rates were found by the vapour transport mechanism than by the atom migration through surface diffusion. Since the processes of vapour phase diffusion and substrate diffusion occur in parallel, the faster of the two will determine the rate of average particle growth. Therefore vapour phase processes need not be considered further. However, one advantage in considering this process is that all the quantities in the theoretical development may be reliably estimated. It does therefore provide a sound basis on which to compare any further theoretical developments.

4.5 General summary of current sintering knowledge

The importance of particle coalescence (which is an important feature of Ruckenstein and Pulvermacher's particle migration model) has been examined by Wynblatt and Gjostein (1975). It has already been shown in Section 4.4.5 that small particles (less than about 20 nm diameter) can migrate over the surface of a typical platinum on alumina catalyst, so there is a finite probability that such particles will collide and coalesce. Therefore sintering rates by the particle migration process may be controlled by either the migration or the coalescence steps. Calculations for the necessary relaxation times for the coalescence process have been made by Wynblatt and Gjostein for various particle sizes; the results obtained are shown in Table 4.7. A comparison of the time values in this table with the time used in Table 4.6 shows that for small particles (less than 50 nm) the coalescence times are very short (seconds) compared with the times required for the particle to migrate a distance equivalent to the interparticle spacing, which is of approximately hours duration. Thus it appears that, for the interparticle spacings of interest in supported catalyst systems, particle migration on the surface is the rate controlling step in the migration–coalescence sequence.

Table 4.7 Relaxation times for particle coalescence at 600°C.

t (s)	d (nm)
1.3×10^{-3}	5
0.3	10
13	25
1.3×10^5	250

The overall picture of sintering of supported metal catalysts seems to have the following pattern. For very small particles (diameter less than about 10–20 nm for platinum) growth occurs predominantly by particle migration, with particle diffusion rate controlling. For larger particles, growth occurs by atom migration on the surface. Vapour transport seems unlikely to be a dominant transport process with platinum but may be of importance for other metals.

Although the experimental data reported in the literature are heavily biased towards platinum catalysts, recent work by Kuo *et al.* (1980) on silica supported nickel catalysts seems to confirm the conclusion noted above. Fitting their data to a sintering power law gave exponent values of 13 and 14 for sintering in nitrogen and hydrogen respectively, at temperatures below 700°C. At 800°C sintering became much faster and the sintering exponent changed to 6 in nitrogen and 4 in hydrogen. The large values of the sintering exponent and changes in particle size distribution during sintering below 700°C suggest a particle migration mechanism, while at 800°C an atomic migration scheme seems to predominate.

Much discussion has ranged on the particle size distributions obtained during sintering. However, for accurate comparisons a large number of particles have to be included in the analysis. This has not always been done, so some of the conclusions reached by this method are suspect.

Ruckenstein and Pulvermacher (1973a,b) show that the binary collision model with particle migration leads to a given sintering exponent. It was shown that the exponent n is related to exponent q of the term $(a/d)^q$ in the equation:

$$D_p = AD_a(a/d)^q \qquad (4.42)$$

by the relation

$$n = q + 4 \qquad (4.43)$$

For spherical particles equation (4.42) becomes (4.41) with $q = 4$, so $n = 8$. For faceted particles, however, q varies with particle size and the value of q rises rapidly above 5 as the particle diameter rises above 2 nm, giving the high values of exponent n already noted.

Since the surface area/particle growth equations (4.7) to (4.10) are interconvertible in terms of exponents n and m, the various sintering processes and resultant exponents may be summarized as in Table 4.8.

Sintering by mechanisms 1–4 is expected to be slower, the greater the cohesive energy of the metal, so the increasing order of stability for supported metals is:

Ag < Cu < Au < Pd < Fe < Ni < Co < Pt < Rh < Ru < Ir < Os < Re

If vapour transport using an oxygen species is dominant, the equilibrium

Table 4.8 Rate power laws for various sintering processes.

Mechanism	Sintering law exponents	
	n	m
1. Spherical particle migration (diffusion control)	8	6
2. Faceted particle migration (diffusion control)	8	6
3. Atomic migration (diffusion control)	5	3
4. Atomic migration (edge control)	4	2
5. Atomic migration (vapour control)	2	1

pressure of the latter is important and the stability sequence would now be expected to be:

$$Os < Ru < Ir < Pt < Pd \approx Rh$$

One vital factor not hitherto discussed in any detail is the surface morphology of the support usually present in industrial catalysts. Also the "wetting" of the support by the metal particle may be important. Surface morphology has two effects: (a) that due to the pore structure of the support; (b) that caused by irregularities on the surface itself leading to energetically stable and metastable positions for the crystallite on the surface. A particle present in a pore is not able to migrate and thus the metal is stabilized. Particle growth will then occur by atomic migration or vapour transport. In either case the particle will grow until it protrudes from the pore (assuming it can fill the pore); then any further growth will cause a smaller value of the particle radius to develop and it will shrink back into the pore.

Increased wetting will also reduce particle growth by the particle migration–coalescence mechanism. This is a result of the increased particle size resulting from the wetting of the support by the particle. Although particle dispersion can be improved by "wetting", most pure metal–ceramic substrate systems do not show significant "wetting" ability.

References

Aben, P. C. (1968). *J. Catalysis* **10**, 224.
Anderson, J. R. (1975). *Structure of Metallic Catalysts*. Academic Press.
Bassett, G. A. (1960). In *Proc. Europ. Regional Conf. on Electron Microscopy* (ed. A. L. Hauwick and B. J. Spit), **1** Nederlandse Vereniging voor Electronen Microscopie, Delft.

Bassett, G. A. (1964). In *Proc. Int. Symp. on Condensation and Evaporation of Solids* (ed. E. Rutner, P. Goldfinger and J. P. Hirth). Gordon and Breach, New York.
Bett, J. A., Kinoshita, R. and Stonehart, P. (1974). *J. Catal.* **35**, 307.
Benesi, H. A., Curtis, R. H. and Studer, H. P. (1968). *J. Catal.* **10**, 328.
Carter, J. L., Casumano, J. A. and Sinfelt, J. H. (1968). *J. Phys. Chem.* **70**, 2257.
Catalyst Handbook (1970). Wolfe Scientific Books.
Chandrasekhar, S. (1943). *Rev. Mod. Phys.* **15**, 1.
Dorling, T. A. (1970). Warren Spring Laboratory, Report LR145(CA). Dept. of Trade and Industry, London.
Flynn, P. C. and Wanke, S. E. (1974a). *J. Catal.* **34**, 390.
Flynn, P. A. and Wanke, S. E. (1974b). *J. Catal.* **34**, 400.
Geus, J. W. (1971). In *Chemisorption and Reactions on Metal Films* (ed. J. R. Anderson), Ch. 3. Academic Press.
Gruber, E. E. (1967). *J. Appl. Phys.* **38**, 243.
Gruber, H. L. (1962a). *J. Phys. Chem.* **66**, 48.
Gruber, H. L. (1962b). *Anal. Chem.* **34**, 1828.
Hermann, R. A., Adler, S. F., Goldstein, M. S. and DeBaun, R. M. (1961). *J. Phys. Chem.* **65**, 2189.
Huang, F. H. and Li, C. (1973). *Scr. Metall.* **7**, 1239.
Hughes, T. R., Houston, R. J. and Sieg, R. P. (1962). *Ind. Eng. Chem. (Proc. Des. Devel.)* **1**, 96.
Jaworska-Galas, Z. and Wrzyszes, J. (1966). *Int. Chem. Eng.* **6**, 604.
Kuo, H. K., Ganesan, P. and DeAngelis, R. J. (1980). *J. Catal.* **64**, 303.
Maat, H. J. and Moscou, L. (1965). *Proc. 3rd Int. Congr. on Catalysis*, p. 1277. North-Holland, Amsterdam.
McLean, M. and Hondros, E. D. (1971). *J. Mater. Sci.* **6**, 19.
Netherlands Patent Application (1968). 6,614,074 (*Chem. Abstr.* **68**, 31814b).
Phillips, W. B., Deslodge, E. A. and Skofronick, J. G. (1968). *J. Appl. Phys.* **39**, 3210.
Renouprez, A., Hoang-Van, C. and Compagnon, P. A. (1974). *J. Catal.* **34**, 411.
Ruckenstein, E. and Dadyburjor, D. B. (1977). *J. Catal.* **48**, 73.
Ruckenstein, E. and Pulvermacher, B. (1973a). *A. I. Ch. E. J.* **19**, 356.
Ruckenstein, E. and Pulvermacher, B. (1973b). *J. Catal.* **29**, 224.
Sagert, N. H. and Ponteau, R. M. L. (1971). *Can. J. Chem.* **49**, 3411.
Satterfield, C. N. (1970). *Mass Transfer in Heterogeneous Catalysis*. M.I.T. Press, Cambridge, Massachusetts.
Skofronick, J. G. and Phillips, W. R. (1967). *J. Appl. Phys.* **38**, 4791.
Smoluchovski, M. V. (1917). *Z. Phys. Chem.* **92**, 129.
Somoraj, G. A. (1968). In *X-ray and Electron Methods of Analysis* (ed. H. von Olphen and W. Parrish), Ch. 6. Plenum Press, New York.
Wanke, S. E. and Flynn, P. C. (1975). *Cat. Rev. Sci. Eng.* **12**, 93.
Whyte, T. E., Kirklin, P. W., Gould, R. A. and Heinemann, H. (1972). *J. Catal.* **25**, 407.
Williams, A., Butler, G. A. and Hammonds, J. (1972). *J. Catal.* **24**, 352.
Wilson, G. R. and Hall, W. K. (1970). *J. Catal.* **17**, 190.
Wynblatt, P. and Gjostein, N. A. (1973). *Scr. Metall.* **7**, 969.
Wynblatt, P. and Gjostein, N. A. (1975). *Progr. Solid State Chem.* **9**, 21.
Wynblatt, P., Della Betta, R. A. and Gjostein, N. A. (1975). In *The Physical Basis for Heterogeneous Catalysis* (ed. E. Drouglis and R. I. Jaffe), p. 501 (9th Battelle Institute Materials Science Colloquia, Gstaad). Plenum Press.
Yates, D. J. C. and Sinfelt, J. H. (1967). *J. Catal.* **8**, 348.

5
Catalyst Deactivation by Poisoning

In this chapter the term poisoning will refer strictly to catalyst deactivation by adsorption of some impurity in the feed stream, or to adsorption of a product of the reaction onto the active sites of the catalyst. Some common poisoning processes are first described, followed by a discussion of means whereby the effects of poisoning may be reduced. This is followed by analyses of the poisoning processes.

5.1 Some common poisoning processes

The most usual type of catalyst poisoning is caused by an impurity that is either present in the gas stream or is formed by some process during the reaction. In both cases this poison becomes adsorbed on the active sites of the catalyst, causing a fall-off in the activity of the catalyst. The poisoning may be either temporary or permanent. The former implies that the poisoning process is reversible and catalyst activity may be restored by removing the source of the poison. In practice it may take some time for the original catalyst activity to be attained but usually this is achieved (or nearly so). Permanent poisoning is, however, a major problem and very often regeneration of the catalyst may not be possible. In these circumstances the catalyst has ultimately to be discarded. Therefore, when permanent poisoning is likely to occur, either the gas stream must be carefully purified or, if this is not possible, the catalytic reactor must be designed in such a way that an economic life may be realized for a given catalyst loading.

Individual examples of catalyst poisoning, both reversible and permanent, are given in the following paragraphs. For convenience these are listed under the process considered, rather than under the actual poison, since a particular poison may deactivate several catalysts.

5.1.1 Steam reforming catalysts

These catalysts are usually in the form of nickel supported on alumina and are extremely sensitive to even the lowest concentrations of poisons.

Elements most frequently encountered as poisons include sulphur, arsenic, halogens, phosphorus, and lead.

(a) *Sulphur*

Sulphur is probably the classic poison for this type of catalyst. It is usually present as impurities in the feedstocks of either natural gas or naphtha fractions used in steam reforming. In natural gas it usually exists as H_2S where concentrations are generally low and less than 300 ppm, except in some sour gas streams, while in naphtha fractions concentrations of organic sulphide of up to 1500 ppm may occur. It appears that all sulphur compounds are readily converted into H_2S over a nickel catalyst at reforming temperatures.

The loss in activity of nickel catalysts due to small amounts of sulphur is greatest for modern more active catalysts. Typically for such catalysts the sulphur concentration must be below 0.5 ppm (Morita and Inoue, 1965; Pichler, 1951). Recent work (Catalyst Handbook, 1970) shows that the poisoning effect of sulphur on nickel reforming catalysts is reversible, the catalytic activity being fully regained when the sulphur concentration in the feed is reduced below the critical level. The sensitivity to sulphur poisoning is increased if the catalysts are operated at lower temperatures. These results, obtained by ICI, and those referred to above are given in Table 5.1. The high results obtained by Pichler suggest that the catalysts used by him were inactive since, as indicated above, larger amounts of sulphur containing poisons are required when the catalytic activity is low.

Poisoning of a nickel catalyst must be associated with reaction between sulphur and the active nickel surface. Since only small concentrations of sulphur are required to poison the catalyst, the formation of bulk sulphide by the reaction $3Ni + 2H_2S = Ni_3S_2 + 2H_2$ is not involved. The amounts of nickel and of sulphur that react are very small. A reforming catalyst containing 15% Ni at 775°C is poisoned when it contains only 0.005% of S, which corresponds to the sulphiding of only 0.06% of the nickel.

Table 5.1 Tolerable sulphur concentrations in the steam reforming of methane.

Catalyst	Reactor exit temp. (°C)			Comments
	800	850	900	
ICI (14–25% Ni)	0.7 ppm	3.5 ppm	17.5 ppm	Reformer inlet temp. 400°C
Morita and Inoue (25% Ni)	2.8	11	—	Isothermal beds
(5% Ni)	—	1.4	—	
Pichler (various catalysts)	42	53	84	Not defined

(b) *Arsenic*

This is effective as a catalyst poison for nickel steam reforming catalysts when the As_2O_3 content in the catalyst exceeds 50 ppm. For practical purposes arsenic poisoning is permanent. Arsenic present in any concentration will accumulate on the catalyst until it produces a detectable effect. Simple steaming of catalyst tubes does not remove the poison.

(c) *Other poisons*

Chlorine and other halogens have an effect similar to sulphur and have about the same concentration limits. As for sulphur, the effect of chlorine and chlorides is reversible.

Copper and lead also deactivate nickel reforming catalysts. Concentrations of lead up to about 3 ppm can be tolerated for short periods of up to a few days, after which, as in the case of arsenic, the accumulation begins to have an effect.

5.1.2 Low-temperature shift catalysts

Copper based catalysts are frequently used to convert carbon monoxide in ammonia synthesis gas streams into carbon dioxide and hydrogen by means of the water gas shift reaction. Copper catalysts are very active for this reaction and can achieve conversions at lower temperatures than the Fe_3O_4/Cr_2O_3 catalysts (high temperature shift catalysts). For the low temperature shift catalysts the most common cause of poisoning is chlorine although, since the active constituent of the catalyst (copper) is present in the form of reduced metal, sulphur poisoning can also occur. Chloride adsorption levels as low as 0.05% can cause loss in activity. This corresponds to a chloride level in the gas phase of 0.001–0.003 ppm, showing that very high degrees of purification are necessary if poisoning is to be avoided completely.

Some interesting results have been reported recently for poisoning of this type of catalyst (Denny and Twigg, 1980). Poison profiles in the bed are steep, suggesting that the rates of poison adsorption may be diffusion limited. Examination of individual catalyst pellets confirms this, since the poison is restricted to a well defined surface zone. Additional evidence is obtained from recent plant data using smaller catalyst pellets (Lundberg, 1979). Catalyst lives are increased when smaller pellets are used because the amount of chloride adsorbed per unit volume of catalyst bed increases as the pellet size decreases, and therefore the rate of movement of the poisoned zone through the bed is slower.

5.1.3 Methanation catalysts

Methanation catalysts have traditionally been employed to clean up small amounts of oxides of carbon in process gas streams, but more recently methanation has become important in its own right in connection with the production of methane from coal processing plants. Usually the catalyst is nickel supported on alumina. Sulphur is a poison for this catalyst as for the similar based nickel reforming catalyst. Arsenic too is a poison. Both of these poisons can be present due to plant upsets, and serious deactivation can occur if the sulphur content is greater than about 0.1–0.2%. Halides also poison this catalyst.

In general, the rate of poisoning is slow with sulphur or arsenic at the levels normally encountered. The activity of the catalyst is affected but not usually the selectivity, so performance can often be improved by raising the operating temperature.

5.1.4 Oxidation catalysts for formaldehyde production

The silver catalysts used for oxidation of methanol to formaldehyde have relatively short lives (less than 1 year on stream). For short lived catalysts the reason for a short time on stream may not be wholly due to activity loss but other factors may also contribute. These may include loss of selectivity or increased pressure drop caused by attrition of the catalyst particles. In the case of silver catalysts used for formaldehyde production, the selectivity is reduced by contamination with the traces of transition metals such as iron, present in the process air. Overall activity remains high and the total conversion of methanol is unchanged, but of course the product distribution is changed towards a higher proportion of unwanted products.

The distinction between selectivity and total conversion is important and is also illustrated by the short life of ammonia oxidation catalysts used in nitric acid manufacture. The platinum/rhodium gauze catalyst used for this reaction shows a selectivity drop from 97 to 95% with increased nitrogen production at the expense of nitric oxide. This decreased selectivity is attributed to increased gas–solid contacting due to increased roughness of the gauze with use. The loss in selectivity, although small, is sufficient to warrant replacement of the gauzes on economic grounds.

5.1.5 Temporary (reversible) poisoning of ammonia catalysts

Ammonia synthesis catalysts are composed primarily of magnetite (Fe_3O_4) with small amounts of potassium, calcium, aluminium, and magnesium oxides. The catalyst suffers temporary poisoning from oxygen and oxygen containing compounds such as H_2O, CO, and CO_2. All of these poisons have

5 Catalyst Deactivation by Poisoning

been found to be equivalent on an oxygen basis, 100 ppm O_2 having the same effect as 200 ppm of H_2O or CO. The poisoning is reversible provided that the duration of the poison addition is not too long. For example a catalyst operating at 450°C and 300 bars shows a 25% reduction in ammonia conversion after 6 days operation with 100 ppm CO added to the feed stream, but activity is completely restored in 1 day when the CO is removed from the feed.

The mechanism of this temporary poisoning is not clear but evidence suggests that the amount of oxygen absorbed by the catalyst is proportional to $(pH_2O/pH_2)^{\frac{1}{2}}$. This points to the poisoning being caused by oxidation of the comparatively small proportion of the iron surface, which is highly active. It can therefore be postulated that the iron surface consists of material with a range of free energies, and the poisoning occurs by oxidation of the catalysts that has the highest free energy. This would be much greater than that of bulk iron.

5.1.6 Reversible poisoning of hydrogenation catalysts

Gioia and coworkers have made an extensive study of the reversible poisoning by water vapour of copper–magnesium catalysts used for ethylene hydrogenation (Gioia and Greco, 1970; Gioia *et al.*, 1970). Similarly, Koh and Hughes (1974) and Burtonwood (1979) observed that ethylene hydrogenation over a supported nickel catalyst could be reversibly poisoned by small quantities of oxygen. These reactions may be similar in nature to the temporary poisoning observed with ammonia synthesis catalysts, in that both poisons contain oxygen, and probably a proportion of the reduced copper or nickel metal crystallites are oxidized by using these poisons.

5.2 Minimization of poisoning

Minimization of deactivation is an important aspect in the design of catalytic reactors. Stability of a catalyst towards sintering can be achieved by a proper choice of materials, including support materials and spacers to minimize migration on the surface. However, chemical poisoning is harder to avoid because it usually involves impure feeds and it may be necessary to consider the various reaction possibilities for these impure feedstocks when selecting a catalyst.

In principle there are three ways of minimizing the effects of poisoning on catalysts:

(a) Purification of the feed stream to a level at which poisoning effects become unimportant.

(b) Use of guard reactors for selective removal of the poison before the main reactor.
(c) Design of the reactor to minimize poisoning effects.

The classic example of feed purification is naphtha desulphurization for steam reforming plants which is well documented. Since the reforming catalyst is nickel, for which sulphur compounds are highly toxic, a high efficiency of removal is required. The usual method is to hydrogenate the sulphur compounds to H_2S over an appropriate catalyst, and absorb the H_2S in a suitable material from the naphtha–hydrogen stream. The hydrodesulphurization reaction is typically catalyzed by a supported cobalt molybdate catalyst. As with most catalysts operating with hydrocarbons, the cobalt molybdate catalysts can themselves become deactivated by carbon deposition, but they can be regenerated. Permanent poisoning can occur, however, by loss of surface or of molybdenum during the regeneration process, and also from small amounts of arsenic which may be present in the feed.

The requirement for efficient absorption of the H_2S liberated, so that the sum of the H_2S leaving the absorption material and any sulphur compounds not hydrogenated does not exceed 0.5 ppm, is met by using zinc oxide as absorbent. This is preferred since, at a hydrodesulphurization temperature of 370°C, the equilibrium value of H_2S is about 1/100 of the above critical value, and ZnO is much superior to the alternative Fe_3O_4 which has a much higher equilibrium level of H_2S at this temperature.

Guard reactors are employed in a number of catalyst systems. Preferably the guard catalyst should not promote other reactions, and for this reason the same catalyst as is used in the main reactor may be used but in different form. One example of this is the low temperature shift catalyst used in ammonia processing streams and already referred to above. This is poisoned by very small amounts of chlorine in the feed, and it has been found that spent low temperature shift catalyst taken from the exit region of a reactor can perform efficiently in this respect. Special low temperature shift catalysts of high activity have also been developed for use as guard catalysts. If the guard catalyst is made from smaller pellets, the absorption efficiency for chlorine removal will usually increase since now a larger amount of impurity will be removed per unit volume of packed catalyst.

The third way of overcoming impurity poisoning is dependent on the form of the activity profile in the bed. If the reaction zone is confined to a small fraction of the bed length, deactivation will usually occur over this small reactive zone. The reaction will then continue on the next zone of the catalyst bed and this process will continue with time until the whole bed length has become deactivated. Thus use of a sufficiently long bed can often give adequate catalyst life. This method is dependent on the poison profile being steep within the bed; if it is shallow, deactivation can occur over the whole bed length (greater in the inlet region) and the method is no longer applicable. The shape of activity profiles is discussed in more detail in a later chapter which considers the effects of deactivation in fixed bed reactors.

5.3 Poisoning of automobile catalysts

An increasingly important class of catalysts includes those used in treating automobile exhaust gases. The aim is to minimize emission of unburnt hydrocarbons, carbon monoxide, and sulphur and nitrous oxides. The need for exhaust treatment arose as a result of legislation, first in the United States but followed by many other countries. It is probable that lead emission will also be curtailed drastically in the near future. There are two types of catalyst employed to deal with exhaust emissions. The predominant catalyst employed is of the noble metal type, usually platinum supported on alumina. Supported transition metal catalysts such as nickel and copper have been proposed but have generally not been adopted to the same extent as platinum catalysts, in spite of their merit of cheapness. For the noble metal catalysts two types of support have been used: the normal pelleted catalyst contained in a conventional fixed bed arrangement in a cylinder, and the more recent monolith or honeycomb type. The latter is a formed block of catalyst material with narrow straight channels in the direction of flow; this arrangement gives low pressure drops. Usually a ceramic block is used to form the monolith matrix; this is usually inert catalytically, the active material being incorporated on a thin alumina "washcoat" which is used to coat the channels of the monolith. This washcoat is only 10–15 μm thick and constitutes 5–15% of the total catalyst weight. The platinum or other active ingredient may be incorporated on the washcoat in the usual manner.

The poisons that may affect automobile catalysts are those arising from constituents or reaction products of the feed, or the lubricating oil, and from other causes (such as construction materials of the exhaust system, including iron, nickel, chromium, and copper). The principal poison in the fuel is the lead additive, together with lead scavengers such as ethylene dibromide and ethylene dichloride which transport the lead additive into the exhaust system. The other major poison in the fuel is sulphur, which may be present to the extent of 0.01–0.1% in the exhaust gases. The principal poison in the oil is phosphorus resulting from the breakdown of oil additives, although some sulphur may also originate from this source.

Early work on the effect of tetraethyl lead additives in gasoline on the poisoning of V_2O_5–CuO–Pd/alumina catalysts was carried out by Weaver

(1969) who first noticed that a large fraction of the total lead passed through the catalyst when high lead levels were present in the fuel. Thus a specificity is present. Later work has confirmed that lead retention is between 13 and 30% and the maximum retention occurs at 850°C. There is not much information currently on the retention of phosphorus but sulphur is retained only by the alumina support of supported platinum catalysts. The deactivation by SO_2 of base metal oxide catalysts used for CO reduction was studied by Jensen *et al.* (1976) and it was observed that, for this particular case, the deactivation was intermediate between homogeneous and pore mouth poisoning. A model developed to account for this type of deactivation predicted particle size effects very well.

The poison distribution has usually been measured on monoliths since these are much easier to sample (by sectioning) than pelleted catalysts. Usually deposition is greatest at the beginning of the bed, but with increasing operating temperature the profile can be flattened. The lead distribution on the catalyst has been found to be affected by the flow pattern particularly in monolith channels, the transition from laminar to turbulent flow favouring lead deposition. Differences between monoliths and pellets have also been observed in terms of contaminant distribution. With pellets very steep gradients for lead deposits have been observed (Klimisch *et al.*, 1975), 65% of the deposited lead being on the external surface of the pellets. However, with monoliths there is a tendency for the deposits to accumulate at cracks in the washcoat, i.e. at the corners of the monolith.

Mass transfer effects due to poisoning have also been investigated. Pore plugging has been shown to be unimportant unless large quantities of lead are present. For phosphorus poisoning Acres *et al.* (1975) claim that phosphorus exhibits pore mouth poisoning whereas lead gives uniform poisoning, but others disagree (e.g. Klimisch *et al.*, 1975). The general consensus now is that deactivation occurs mainly by pore mouth poisoning.

The legislative impetus given to this class of catalysts has resulted in intensive investigations of poisoning relevant to other catalyst applications. The literature has been well reviewed on this subject, one example being that of Shelef *et al.* (1978).

5.4 Analysis of poisoning processes

5.4.1 Irreversible poisoning

Following the general analyses of Wheeler (1951), one of the earliest treatments of the problem of poisoning by an impurity in the feed was given by Masamune and Smith (1966). In their paper they undertook a study of the fouling of catalyst pellets by parallel and series mechanisms. These are

5 Catalyst Deactivation by Poisoning

discussed in Chapter 6 when fouling is considered. However, also included in the results was an analysis of fouling by an impurity in the feed. Within the classification adopted in this text, this falls into the general area of chemical poisoning by a feed impurity. Because the basic aim of Masamune and Smith's work was to analyze fouling, the procedures used were adapted to conform to the analysis of fouling phenomena. Thus it was assumed that deactivation occurred by deposition of material blocking the active sites, and a simple linear relation was assumed between the rate of removal of the poisoning impurity and the active surface remaining. This kind of poisoning, designated "independent fouling" by Masamune and Smith, is mathematically similar to first order fouling by a parallel mechanism (q.v.). Further assumptions used in the model equations include those of isothermality and the absence of external film resistances.

The main reaction is assumed to be of the following form:

$$A(g) \rightarrow B(g) \tag{5.1}$$

Deactivation is caused by deposition of a gaseous impurity, P, present in the feed stream. The concentration of deposit P as a function of time is given by:

$$\frac{dq}{dt} = k_P(1 - \psi)C_P \tag{5.2}$$

where ψ is equal to q/q_o with q_o the concentration of P on the surface when deactivation is complete, i.e. the poisoning is linear in surface coverage by the poison. The terms k_P and C_P represent the rate constant for impurity poisoning and the gaseous concentration of poison respectively.

Mass balances for the main reactant, A, and impurity poison, P, in a single spherical pellet may be written as follows:

$$D_{eA}\nabla^2 C_A - \varepsilon\frac{\partial C_A}{\partial t} - \rho k_A C_A(1 - \psi) = 0 \tag{5.3}$$

$$D_{eP}\nabla^2 C_P - \varepsilon\frac{\partial C_P}{\partial t} - \rho k_P C_P(1 - \psi) = 0 \tag{5.4}$$

Boundary and initial conditions for the above system of equations are:

$$\psi = 0; \quad t = 0; \quad r_o \geqslant r \geqslant 0 \tag{5.5}$$

$$C_A = C_{Ao}; \quad C_P = C_{Po}; \quad t \geqslant 0, r = r_o \tag{5.6}$$

$$\frac{\partial C_A}{\partial r} = \frac{\partial C_P}{\partial r} = 0; \quad r = 0, t \geqslant 0 \tag{5.7}$$

It is assumed that the diffusivities D_{eA} and D_{eP} are independent of the concentration q deposited on the catalyst surface and also independent of gas composition. Furthermore, the time required to reach steady state with respect to the accumulation of mass in the void space of the catalyst pellet is very short compared with the time in which catalyst activity changes significantly. Therefore the accumulation terms in equations (5.3) and (5.4) can be neglected.

The above mass balances for A and P and equation (5.2) together with the appropriate boundary and initial conditions are then transformed into dimensionless expressions, using the dimensionless variables $\delta = r/r_o$, $a = C_A/C_{Ao}$, and $p = C_P/C_{Po}$. The balance for reactant A then becomes:

$$\frac{1}{\delta^2}\frac{d}{d\delta}\left(\delta^2 \frac{da}{d\delta}\right) - \phi^2(1 - \psi)a = 0 \tag{5.8}$$

The corresponding balance for P is given by:

$$\frac{1}{\delta^2}\frac{d}{d\delta}\left(\delta^2 \frac{dp}{d\delta}\right) - \phi_P^2(1 - \psi)p = 0 \tag{5.9}$$

and

$$\frac{d\psi}{d\tau} = (1 - \psi)p \tag{5.10}$$

where ϕ_P is the Thiele modulus defined in terms of the poison rate constant and diffusivity, and τ the dimensionless time is defined as:

$$\tau = \frac{k_P C_{Po} t}{q_o} \tag{5.11}$$

Equations (5.8), (5.9), and (5.11) may be solved for a, p, and the fraction of free area $(1 - \psi)$ available. For this particular case, the poisoning is a function of τ and the two different Thiele moduli ϕ and ϕ_P. Profiles of a, p, and $(1 - \psi)$ were obtained by Masamune and Smith (1966); the results from this paper are given in Fig. 5.1 which shows the general profiles for a somewhat high value (10) of ϕ_P. With this high value of ϕ_P, the diffusion resistance is sufficiently high to give almost complete pore mouth deactivation in the outer regions of the catalyst pellet. This is seen from the lower part of Fig. 5.1 where $(1 - \psi)$ is almost zero as r/r_o approaches unity. For smaller values of ϕ_P these curves could be modified considerably. In many ways the value of ϕ_P considered in the analysis is misleading. For a value of ϕ_P equal to 10, the rate constant for the deactivation reaction would be too large to allow for the pseudo-steady state assumed, and/or the diffusivity of the impurity poison would have to be very small. However, the analysis is important in showing the general profiles that might be expected under certain conditions.

Effectiveness factors were also obtained as a function of time for impurity poisoning by Masamune and Smith. These showed a decrease with time on

5 Catalyst Deactivation by Poisoning

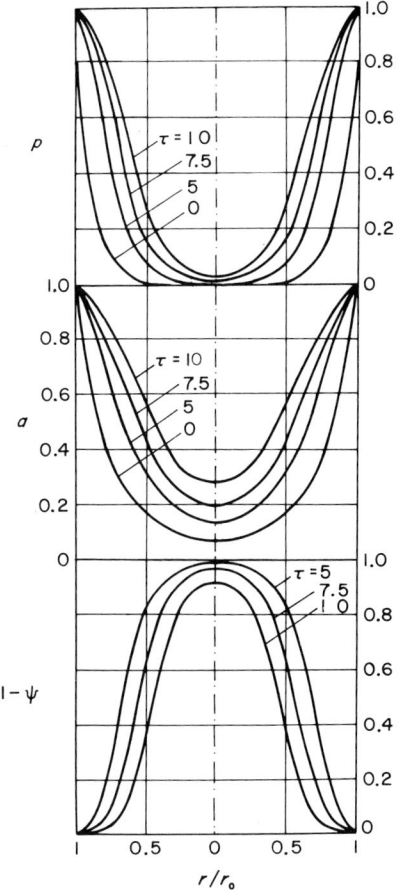

Fig. 5.1 Profiles for independent poisoning; $\phi = 5$, $\phi_P = 10$. (Masamune and Smith, 1966)

stream, as would be expected, the extent of the decrease increasing considerably with increasing Thiele modulus for the main reaction. As the intraparticle resistance for the poison increases, less of the catalyst interior should be deactivated significantly by the poison. The quantitative effect of this is seen in Fig. 5.2. Here, effectiveness factors are compared as a function of time for increasing ϕ values for the same value of ϕ_P (10). The dashed line on this figure represents the lower ($\phi = 0$) limit of the diffusional resistance for the main reaction. These curves demonstrate that there is a continual decrease in effectiveness factor with increase in diffusion resistance. If the intraparticle diffusion resistance for the poison increases, less of the catalyst interior should

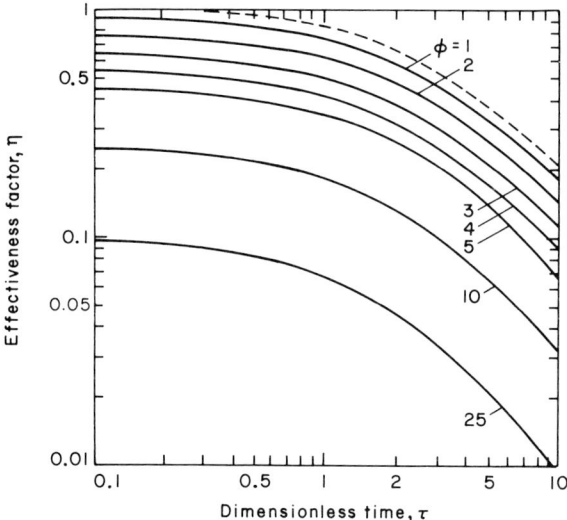

Fig. 5.2 Effectiveness factor for independent poisoning; $\phi_P = 10$. (Masamune and Smith, 1966)

be deactivated by the poison. These results demonstrate that, for poisoning under these conditions, least deactivation occurs when there is a minimum diffusion resistance for the main reaction and a maximum resistance to diffusion for the poison.

A more detailed analysis of impurity poisoning was given by Hegedus (1974). The restriction that external film resistances were absent was now removed, although isothermal conditions were still assumed. Irreversible poisoning was assumed to be the deactivation process with all active sites on the catalyst surface being assumed equivalent, i.e. each site could adsorb either reactant or poison equally well. Both the main reaction and the poisoning reaction were written in a general non-linear form. Thus, for the main reaction, A → B, the rate of decrease of reactant concentration is given by:

$$-\frac{dC_A}{dt} = k(\theta_o - \theta_P)^m C_A^n \qquad (5.12)$$

where θ_o is the initial area of the active sites and θ_P is the poisoned area of the active sites.

The reaction of the poison, P, with the sites is represented by a non-linear irreversible process, P → W, where W represents the poison adsorbed on the

5 Catalyst Deactivation by Poisoning

active sites. The rate of increase of adsorbed poison concentration is then:

$$\frac{dC_W}{dt} = k_P(\theta_o - \theta_P)^p C_P^q \tag{5.13}$$

Usually in poisoning processes, as already seen, the time scale of the poisoning reaction is very much greater than that of the main reaction. It follows that the rate of accumulation of species A and P is negligible and, if the diffusivities are also independent of concentration steady state, balances may be written for A and P in the pellet as was done for equations (5.3) and (5.4) previously. These balances may then be written as:

$$D_{eA}\nabla^2 C_A - k(\theta_o - \theta_P)^m C_A^n = 0 \tag{5.14}$$

$$D_{eP}\nabla^2 C_P - k_P(\theta_o - \theta_P)^p C_P^q = 0 \tag{5.15}$$

Equations (5.14) and (5.15) together with the rate expression (5.13) for adsorbed poison concentration can be solved with appropriate boundary conditions to give concentration profiles for both reactant and poison in a single catalyst pellet at various times. Suitable boundary conditions which account for the possibility of a finite external film mass transfer resistance are:

$$\left.\begin{array}{l} C_A = f(r) \quad \text{at } t = 0, \text{ any } r \\ C_P = 0 \quad \text{at } t = 0, \text{ any } r \\ \theta_P = 0 \quad \text{at } t = 0, \text{ any } r \end{array}\right\} \tag{5.16}$$

$$\left.\begin{array}{l} \dfrac{dC_A}{dr} = 0, r = 0, t \geqslant 0 \\[2mm] \dfrac{dC_P}{dr} = 0, r = 0, t \geqslant 0 \end{array}\right\} \tag{5.17}$$

$$\left.\begin{array}{l} D_{eA}\dfrac{dC_A}{dr}\bigg|_{r=R} = k_{cA}(C_{Ao} - C_A|_{r=R}) \\[2mm] D_{eP}\dfrac{dC_P}{dr}\bigg|_{r=R} = k_{cP}(C_{Po} - C_P|_{r=R}) \end{array}\right\} \tag{5.18}$$

where k_{cA} and k_{cP} are the appropriate mass transfer coefficients.

These equations may be made dimensionless in the usual way by defining dimensionless variables for C_A and C_P. Additional parameters that are then required include the Thiele moduli and Biot numbers for reactant and poison species respectively:

$$\phi_A = R\sqrt{\frac{kC_{Ao}^{n-1}\theta_o^m}{D_{eA}}} \quad (5.19)$$

$$\phi_P = R\sqrt{\frac{k_P C_{Po}^{q-1}\theta_o^p}{D_{eP}}} \quad (5.20)$$

$$\text{Sh}_A^* = \frac{Rk_{cA}}{D_{eA}} \quad (5.21)$$

$$\text{Sh}_P^* = \frac{Rk_{cP}}{D_{eP}} \quad (5.22)$$

In the above relations when ϕ_A is less than unity this implies, as usual, the absence of internal diffusional resistances, while for $\phi_A > 1$ these will be present. On the other hand, a ϕ_P value less than unity implies uniform poisoning while $\phi_P > 1$ signifies that strong pore mouth poisoning effects will be present. Additionally, values of the modified Sherwood numbers less than unity indicate strong external mass transfer control, while $\text{Sh}^* > 1$ indicates that these external mass transfer effects are less important.

Hegedus (1974) investigated combinations of varying degrees of external and internal mass transfer control. A typical example of his results, for the case of strong internal and weak external mass transfer control, is given in Fig. 5.3 where concentration profiles for reactant and poison as well as the

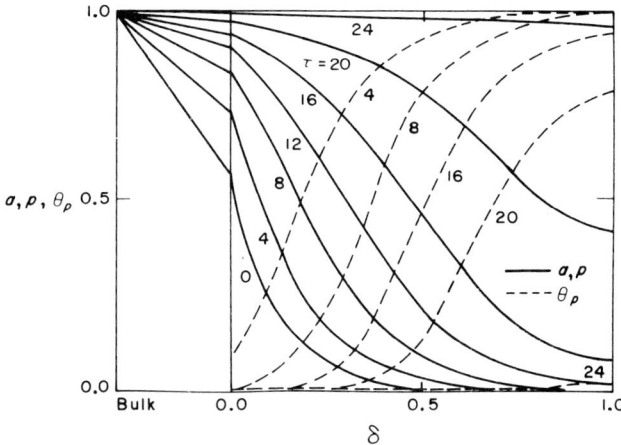

Fig. 5.3 Distribution of a, p, and θ in a spherical catalyst pellet; strong internal, weak external mass transfer control. (Hegedus, 1974)

5 Catalyst Deactivation by Poisoning

fraction of surface unpoisoned are given for a spherical catalyst pellet. For this particular case, $\phi_A = \phi_P = 10$ and $Sh_A^* = Sh_P^* = 10$. Both the main and poisoning reactions are assumed to be first order with respect to the appropriate gaseous reactant and the fraction of active sites. The dimensionless time is the parameter on these curves. An important feature of this figure is that it demonstrates that, as poisoning (i.e. time) increases, the reactant concentration within the pellet also increases, showing that less reactant is consumed by the main reaction.

A vital question, frequently posed, concerns the effect of internal and external mass transfer resistances on the poisoning behaviour of the pellet. Frequently the problem is to estimate the effective lifetime of the catalyst under various conditions. To do this it is convenient to plot the relative overall reaction rate for the pellet at a particular time, $\bar{R}/\bar{R}_o$, against the dimensionless time. Here $\bar{R}_o$ is the initial overall reaction rate, i.e. the reaction rate with no poison present. The effect of various combinations of parameters was investigated by Hegedus (1974) and the results obtained are given in Fig. 5.4.

Curves 1 and 2 compare the effects of increasing the internal diffusional resistances (ϕ goes from 1 to 10) while decreasing the external resistances (Sh* increases from 1 to 10). An increase in the internal resistance is seen to be

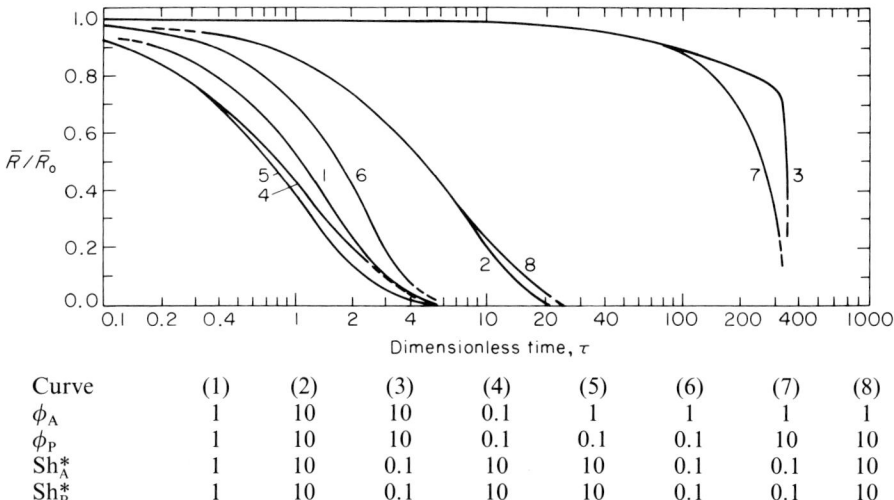

Curve	(1)	(2)	(3)	(4)	(5)	(6)	(7)	(8)
ϕ_A	1	10	10	0.1	1	1	1	1
ϕ_P	1	10	10	0.1	0.1	0.1	10	10
Sh_A^*	1	10	0.1	10	10	0.1	0.1	10
Sh_P^*	1	10	0.1	10	10	0.1	0.1	10

Fig. 5.4 Time-dependent decay of the activity in spherical catalyst pellets ($n = m = p = q = 1$). The lifetime of the catalyst increases with increasing internal and external transport resistances. Note that the initial effectiveness factor decreases in that direction. (Hegedus, 1974)

predominant in extending the life of the catalyst pellet. The most favourable extension of catalyst life is seen to occur for curve 3 where both internal and external resistances are increased, and also for curve 7 where, although the internal diffusion resistance to the main reaction is small, a large value of ϕ_P plus strong external resistances ($Sh_A^* = Sh_P^* = 0.1$) gives an extended catalyst life. However, it should be pointed out that the increased catalyst life is not obtained without other effects. Increase of mass transfer resistances results in a decrease in effectiveness factor and thus a much longer reactor is required to achieve the same overall conversion. It should also be noted that, in order to make the comparisons illustrated in Fig. 5.4, catalysts are compared at the same temperature using the same active component at the same poison concentration under identical catalyst loadings.

More recently, Hegedus and McCabe (1981) used a more elaborate system of equations to model irreversible poisoning on a catalyst pellet. Essentially, a time dependent analysis is derived, assuming an isothermal first order conversion of a reactant species A, poisoned by a species W which results from an irreversible single site reaction of P on an individual site. Therefore the chemical reactions may be written:

$$A + S \rightleftharpoons AS$$

$$AS \rightleftharpoons B + S$$

$$S + P \rightleftharpoons PS$$

$$PS \rightarrow WS$$

In this scheme all sites are assumed to be equivalent and B does not chemisorb.

Transient balances may then be written for species A, P, and W, and the surface fractions θ_A and θ_P. The development assumes that the surface fractions θ_A and θ_P may be expressed in terms of C_A and C_P using equilibrium theory. This is difficult to justify except for the case of slow transients. Successive simplification from the equilibrium Langmuir–Hinshelwood relations, and removal of the time dependent term (which is reasonable for the usual case where the poisoning reaction is relatively slow compared with the main reaction), leads ultimately to the following three equations:

$$D_{eA}\nabla^2 C_A - ak_A K_A C_A(1 - \theta_W) = 0 \tag{5.23}$$

$$D_{eP}\nabla^2 C_P - ak_P K_P C_P(1 - \theta_W) = 0 \tag{5.24}$$

$$\frac{\partial \theta_W}{\partial t} = k_P K_P C_P(1 - \theta_W) \tag{5.25}$$

These three equations are analogous to those adopted in the classic analysis of impurity poisoning by Masamune and Smith (1966) (see above).

5 Catalyst Deactivation by Poisoning

To summarize, it seems that, while there have been considerable developments in interpreting irreversible poisoning processes, there are still considerable gaps in our knowledge. For example, most treatments assume that uniform sites are available for both reactant and poison. Additionally, the non-isothermal case has not been considered in any detail; this is an important omission since most reactions of commercial importance operate non-isothermally.

5.4.2 Reversible poisoning

Reversible poisoning has been shown to be important in many industrial processes, mainly as a means of controlling one or more of the main reactions. In addition to those cited at the beginning of this chapter, others include the addition of chlorinated hydrocarbons to control ethylene oxidation over a silver supported catalyst and hydrocarbon reforming.

The first attempts at analyzing such systems were made by Gioia (1971) and Gioia et al. (1970), based on a mechanism of poisoning proposed by Innes (1954). These investigators also obtained confirmation of their modelling predictions in experimental measurements of the rate of hydrogenation of ethylene over a copper–magnesia catalyst using water as the reversible poison.

The main feature of reversible poisoning is of course that catalytic activity should be attained again after removal of poison from the feed stream. True reversible conditions then hold and a Langmuir mechanism for adsorption will be fully valid. Therefore, for simple adsorption of poison, at adsorption equilibrium:

$$\frac{\theta_P}{\theta_o} = \frac{\bar{C}_P}{\bar{C}_{P\,max}} = \frac{K_P C_P}{1 + K_P C_P} \qquad (5.26)$$

where $\bar{C}_P$ is the adsorbed phase poison concentration and $\bar{C}_{P\,max}$ is the maximum adsorbed phase poison concentration.

The poison penetration may be analyzed for flat slab pellet geometry by use of the relation:

$$D_{eP} \frac{\partial^2 C_P}{\partial x^2} = \frac{\partial C_P}{\partial t} + \frac{\partial \bar{C}_P}{\partial t} \qquad (5.27)$$

with boundary conditions that assume the absence of external film resistances.

Solution of equations (5.26) and (5.27) enables the poison concentration profile in the pellet to be determined. This can only be done numerically, but two limiting cases may be considered which enable some useful generalizations of the problem of reversible poisoning to be made.

The first case is when $K_P C_P \ll 1$ which causes equation (5.26) to simplify to the linear form:

$$\bar{C}_P / \bar{C}_{P\max} = K_P C_P \tag{5.28}$$

Substitution in (5.27) gives a solution for $\bar{C}_P$ in the form:

$$\bar{C}_P / \bar{C}_{P\max} = K_P C_P \, \text{erfc} \, \frac{x}{2\sqrt{\sigma\tau}} \tag{5.29}$$

where

$$\sigma = \frac{D_{eP}}{K+1} \quad \text{and} \quad K = \bar{C}_{P\max} K_P$$

The alternative limit is when $K_P C_P \gg 1$. Then the concentration of adsorbed poison can be described by:

$$\bar{C}_P / \bar{C}_{P\max} = 1 \tag{5.30}$$

which corresponds to the pore mouth poisoning case described by Wheeler with the outer region of the catalyst slab deactivated. This relation is valid up to a distance λ from the catalyst surface, with λ a function of time (λ increases as time increases). For penetration of poison greater than λ, the adsorbed poison concentration is now:

$$\bar{C}_P = K C_P^* \, \text{erfc}(x - \lambda/2\sqrt{\sigma t}) \tag{5.31}$$

where C_P^* is the concentration of poison in the gas phase in equilibrium with the maximum adsorbed poison concentration $\bar{C}_{P\max}$. These two extreme cases are illustrated in Fig. 5.5 and show the pore mouth poisoning effect for the second case quite clearly.

The above poison penetration profile equations may be coupled with a main first order reaction, A → B. If independence of adsorption between reactant and poison is assumed, a mass balance on A may be written:

$$D_{eA} \frac{\partial^2 C_A}{\partial x^2} = \frac{\partial C_A}{\partial t} + k\left(1 - \frac{\theta_P}{\theta_o}\right) C_P \tag{5.32}$$

The transient term in equation (5.32) may be omitted if the usual assumption of a pseudo-steady state is used, i.e. the rate of change of catalyst activity is slow compared with the rate of diffusion of reactant. This leads to the simpler relation:

$$D_{eA} \frac{d^2 C_A}{dx^2} = k\left(1 - \frac{\theta_P}{\theta_o}\right) C_A \tag{5.33}$$

This equation was solved by Gioia (1967) for the semi-infinite state. The

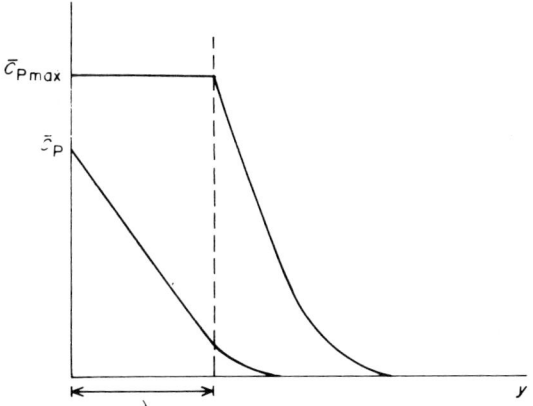

Fig. 5.5 Sketch of poison concentration profile in a slab. (Gioia, 1971)

formal similarity of this expression to the rate part of (5.12) as used by Hegedus (1974) should be noted.

The results obtained by Gioia were, however, related only to the case where the main reaction was not under any form of diffusional limitation, i.e. powder particles, or values of the Thiele modulus less than unity were considered. A more general approach was developed by Valdman et al. (1976) where these and other restrictions (e.g. the semi-infinite geometry of the pellet) were removed. In this analysis by Valdman an isothermal catalyst pellet was assumed in the first instance, together with a uniform effective diffusivity for both reactant and poison which are independent of concentration. The main reaction may be diffusion controlled, and Langmuir adsorption of poison and reactant is assumed. External transport limitations are assumed to be negligible.

Consider a reaction, A → Products, which occurs by a Langmuir–Hinshelwood mechanism, with a reaction rate r. A differential balance for the reacting component A in an isothermal flat slab may be written:

$$D_{eA} \frac{d^2 C_A}{dx^2} = R^* \tag{5.34}$$

Assuming that the poison adsorbs independently of any other component, a mass balance for the poison component P yields:

$$D_{eP} \frac{\partial^2 C_P}{\partial x^2} = \frac{\partial \bar{C}_P}{\partial t} \tag{5.35}$$

where C_P and $\bar{C}_P$ are the gas phase and adsorbed phase concentrations of poison respectively. If the adsorption of poison is rapid, equilibrium conditions prevail and C_P and $\bar{C}_P$ are related by the Langmuir adsorption isotherm:

$$\frac{\bar{C}_P}{\bar{C}_{P\,max}} = \frac{K_P C_P}{1 + K_P C_P} \tag{5.36}$$

where $\bar{C}_{P\,max}$ is the maximum adsorbed phase poison concentration in equilibrium with the gas phase.

When poison is competing with reactant for the same available sites on the catalyst surface, the rate expression is:

$$R^* = \frac{kC_A}{1 + K_A C_A + K_P C_P} \tag{5.37}$$

Equations (5.34) and (5.37) may be combined to give:

$$\frac{dC_A^2}{dX^2} = \phi^2 \frac{C_A'}{1 + K_A C_{Ao} C_A' + K_P C_{Po} C_P'} \tag{5.38}$$

and

$$\frac{\partial C_P'}{\partial X^2} = \frac{1}{(1 + K_P C_{Po} C_P')^2} \frac{\partial C_P'}{\partial \tau} \tag{5.39}$$

where C_A', C_P', X, and τ signify dimensionless concentration of reactant and poison, length, and time respectively, and the subscripts o on the concentration terms are for surface values.

The above equations may be solved simultaneously with the following boundary conditions:

$$X = 0 \quad C_A' = C_P' = 1 \quad \text{for } \tau > 0$$

$$X = 1 \quad \frac{dC_A'}{dX} = \frac{dC_P'}{dX} = 0 \quad \text{for } \tau > 0$$

$$\tau = 0 \quad C_A' = 1, C_P' = 0 \quad \text{for } 0 < X < 1$$

The overall effectiveness factor of a catalyst pellet is characterized by the ratio of the actually observed rate to the reaction rate based on surface conditions in the absence of any poisoning or external mass transfer resistance. For the present case this may be written:

$$\eta = \frac{1 + K_A C_{Ao}}{\phi^2} \frac{dC_A'}{dX}\bigg|_{X=0} \tag{5.40}$$

The surface differential dC_A'/dX is determined by integrating equation (5.38) and taking the value at $X = 0$. As the reaction rate R^* is a function of both reactant and poison, this integration can normally only be done numerically as a two point boundary value problem, except for the special

5 Catalyst Deactivation by Poisoning

case of a uniform poison distribution. Thus a trial and error procedure is necessary. An alternative method of calculating η is to use the orthogonal collocation polynomial approximation (Villadsen and Stewart, 1967) using a single parameter trial function (Valdman et al., 1976). This is usually adequate for most purposes but if greater accuracy is needed, as when steep concentration profiles occur, more collocation points may be employed or the reaction zone collocation method (Kam et al., 1976) may be used.

Limiting values of η may also be obtained using this approach. Valdman et al. (1976) noted that at long reaction times, when the poison can be assumed to be uniformly adsorbed on the catalyst, $C'_P = 1$ and, using a form of Petersen's approximation for large Thiele moduli (Ramachandran, 1975), the effectiveness factor is:

$$\eta = \frac{1 + K_A C_{Ao}}{\phi} \left[\frac{2}{K_A C_{Ao}} \left(1 - \frac{\ln(1 + V)}{V} \right) \right]^{\frac{1}{2}} \qquad (5.41)$$

where $V = K_A C_{Ao}/(1 + K_P C_{Po})$. For unpoisoned catalyst ($C_{Po} = 0$) the above formula is valid for $\phi > 3$. For a poisoned catalyst the value of ϕ corresponding to the beginning of the asymptotic region is larger and depends on the amount of poison adsorbed on the catalyst.

For low values of the Thiele modulus the concentration profile of reactant is flat and C'_A tends to unity. For a uniform distribution of poison η is given by:

$$\eta(\tau = \infty) = \frac{1 + K_A C_{Ao}}{1 + K_A C_{Ao} + K_P C_{Po}} \qquad (5.42)$$

Solutions were obtained by Valdman (1970) for a typical set of conditions, namely $D_{eP} = 10^{-6}$ m^2 s^{-1}, L = pellet half thickness = 6 mm, $K_P = 10^3$ m^3 mol^{-1}, $C_{P\,max} = 10^{-2}$ kmol m^{-3}, and $K_A C'_A = 1$, dimensionless. Use of these values leads to a characteristic time, t^*, for the system equal to 100 h ($t^* = L^2 K_P \bar{C}_{P\,max}/D_{eP}$).

It should be noted in the above that the Thiele modulus, ϕ, refers to the main reactant since this is the parameter that influences the effectiveness of the main reaction. The concentration distributions of reactant for infinite time (i.e. at equilibrium) were compared by Valdman et al. (1976) for two levels of poison concentration characterized by $K_P C_{Po}$ equal to 1 and 10 and for three values of the Thiele modulus equal to 1, 3, and 10. The results are presented in Fig. 5.6 for flat slab geometry. It can be seen that there is a large spread in the concentration profiles at intermediate values of the Thiele modulus ($\phi = 3$). At Thiele moduli equal to 1 and 10 the change in concentration profiles due to poisoning is much less pronounced.

For comparing the effect of size and other characteristics of catalyst pellets it is convenient to plot the variation of effectiveness factor as a function of

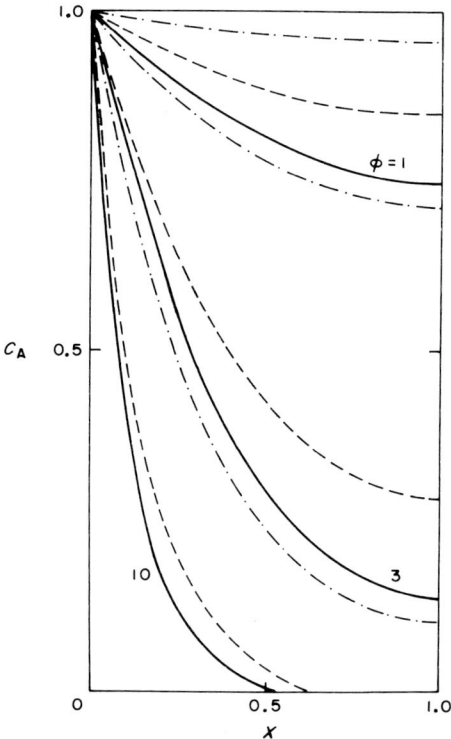

Fig. 5.6 Steady-state distribution of reactant, in a flat slab catalyst pellet, at different levels of poison concentration. The value of $K_P C_{P_o}$ is 0 for the solid curves, 1 for the broken curves, and 10 for the dot-dash curves. (Valdman et al., 1976)

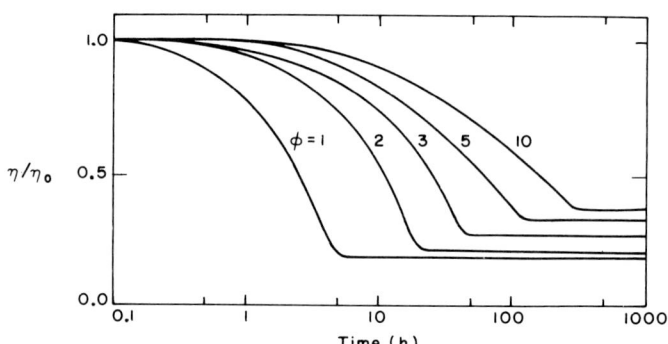

Fig. 5.7 Time-dependent variation of the ratio η/η_o during poisoning, for various values of Thiele moduli, of flat slab catalyst pellets. (Valdman et al., 1976)

real time, t. The ratio η/η_o, which represents the ratio of the effectiveness factor at any time to that for the unpoisoned catalyst, is shown in Fig. 5.7 for five values of the Thiele modulus. The ratio of effectiveness factors denotes the relative drop in reaction rate caused by poisoning as a function of time and is analogous to the plot in Fig. 5.4 for irreversible poisoning. For small values of the Thiele modulus, the poisoning is completed in much less time than for larger values and is also accompanied by a larger relative drop in activity. The reverse is true for larger Thiele moduli. For example, a system with a Thiele modulus of 1 drops to 18% of its original activity at the steady state and this takes less than 10 h, while a system with a Thiele modulus equal to 10 drops only to 37% of its original activity and this takes more than 300 h. A similar trend is also observed at any given time. For example, 3 h after the introduction of poison into the system, the relative drops in activity are 58% and 2% for values of the Thiele modulus equal to 1 and 10 respectively.

Figure 5.7 is useful in estimating the best size of pellet to minimize poisoning. For the conditions chosen for the comparison it would seem that, if particle size alone determines the magnitude of the Thiele modulus, larger particles should have the best resistance to impurity poisoning. However, it should be borne in mind that Fig. 5.7 shows only the *relative* effectiveness factors. Thus, whilst a large Thiele modulus gives the lowest relative decrease in activity, the original activity (without poisoning) may be very low in any case for the larger pellets because of the increased diffusional resistance.

Other factors considered in this same study were the description kinetics and an analysis of reactor performance based on extrapolation of the single pellet results. It was observed that, under the conditions used in the simulation, the time required to recover the original activity when the impurity was removed from the feed stream was much greater than the time required for poisoning. These predictions agree with the experimental data Gioia (1971) for the reversible water vapour poisoning of the ethylene hydrogenation reaction over a copper–magnesia catalyst.

The reactor performance or productivity may be estimated from the area under the curve obtained by plotting effectiveness factor against time. The results obtained by Valdman for both continuous and intermittent poisoning showed that pellets having an intermediate Thiele modulus (~ 2) gave a higher productivity in many cases. This is an interesting result since it is similar to that obtained with isothermal parallel fouling, to be described in Chapter 6.

5.4.3 Non-isothermal effects

As noted above, little work has been done on the coupled effects of poisoning and non-isothermal reaction in a catalyst pellet. Since endothermic processes tend to decrease the overall temperature in the catalyst pellet, this reinforces the poisoning effect and will not be considered further. The more interesting effects occur with exothermic processes where increase in temperature within the pellet can modify the poisoning effect to some extent.

As poisoning action increases with time the rate of the main reaction will diminish, causing the net rate of heat production to diminish, so in the later stages of poisoning near-isothermal conditions may prevail within the pellet. However, the initial pattern of poison adsorption must be influenced by temperature effects within the pellet. Although little previous attention has been given to this point, recent temperature monitoring by two sets of workers has provided some interesting results.

Following on the work of Koh and Hughes (1974), in which a catalyst pellet instrumented with thermocouples was used to record temperatures during the hydrogenation of ethylene, Burtonwood (1979) demonstrated conclusively that reversible oxygen poisoning was occurring in this system. Results for the unpoisoned pellet and a pellet subjected to an oxygen concentration of 0.0345% in the gas stream are illustrated in Fig. 5.8. Only

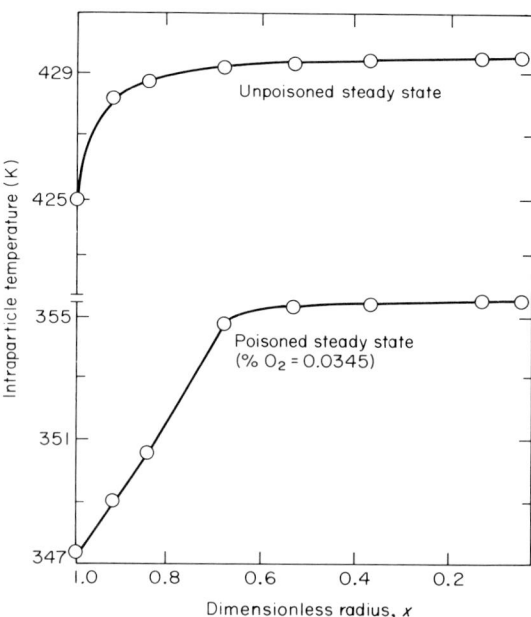

Fig. 5.8 Steady-state intraparticle temperature profiles for the unpoisoned and poisoned steady states; $T_{oi} = 323.7$ K, % $C_2H_4 = 10.2$, $F_T = 8.33 \times 10^{-6}$ m^3 s^{-1}.

the steady state profiles are shown in the figure, for clarity, but the curves do show a reduction of the flat central plateau temperature with time, indicating that the outer regions of the pellet are being progressively deactivated. Some confirmation of this conclusion is also given by the upward concavity of the temperature profile near the pellet surface. This would be expected if the pellet had a central reacting region surrounded by an inactive shell. Also of importance is the relative decrease of the external temperature rise and an enhanced relative intraparticle temperature rise.

Similar quantitative results have been obtained for poisoning by thiophene of the benzene–hydrogen reaction (Butt *et al.*, 1977). They showed very clearly that the relative importance of the intraparticle temperature profile increases as the catalyst becomes progressively poisoned. This is an example of irreversible poisoning.

An increase in temperature may have two effects on a catalytic reaction system subject to poisoning. The first is the normal increase in rate which will be observed for any reaction whether poisoned or not. The second effect is more complex in that a change in temperature will affect the extent of adsorption of both reactant and poison. Most analyses have concentrated on the effect of temperature on the rate alone and have not considered adsorption effects.

Valdman *et al.* (1976) developed their single pellet model to account for non-isothermal behaviour using Dirichlet boundary conditions as for their isothermal case (i.e. external resistances due to mass and heat transfer are assumed to be absent). Adsorption constants for reactant ($K_A C_{Ao}$) and poison ($K_P C_{Po}$) were assumed to be 1 and 10 respectively, and the effects of temperature were considered at three values of the product of thermicity factor, β, and Arrhenius number, γ, and at various times of poisoning.

For small values of the Thiele modulus the distribution of reactant inside the catalyst prior to poisoning is effectively uniform. The presence of a poison under these conditions leads to less reaction of A, making the concentration of A even more uniform. If large particles are employed or if the Thiele modulus is large, reaction in the absence of poisoning is complete within a thin shell next to the surface of the pellet. The presence of poison causes fewer active sites to be available for the reactant, and consequently reactant will penetrate deeper into the pellet to use the non-occupied sites.

The simulated results of Valdman *et al.* (1976) were in the form of plots of effectiveness factor against Thiele modulus and are given in Table 5.2 for a system in which, for the main reaction, $\gamma = 10$ and $\beta = 0.4$. Values of the effectiveness factor, η, at $\phi = 0.3$ occur before the maximum in the η vs. ϕ curves, while those for $\phi = 4$ are after the maximum in the asymptotic region. Inspection of Table 5.2 shows that the drop in activity is large for small values of the Thiele modulus, but the process is relatively slow. This

Deactivation of Catalysts

Table 5.2 Effectiveness factor (η) for a poisoned catalyst pellet at two Thiele moduli; $K_P C_{Po} = 10$, $K_A C_{Ao} = 1$.

Time, τ	η for $\phi = 0.3$	η for $\phi = 0.4$
0	1.0	0.83
0.01	0.8	0.6
0.04	0.6	0.5
∞	0.16	0.3

behaviour is expected since, under these conditions, the active sites are totally used because reactant penetrates throughout the particle. Similarly, the small particle dimensions limit any temperature rise. Thus, under these conditions, the effectiveness factor is a function only of the time of poisoning and shows a large drop from the initial unpoisoned level.

However, for larger values of the Thiele modulus the decrease in effectiveness factor is smaller. This is due to the compensating effect of temperature present at this larger Thiele modulus. Conversely, this increase in Thiele modulus results in a reduced time to reach the poisoned steady state ($\tau = \infty$).

The effect of external resistances of both mass and heat transfer on the effectiveness of a single catalyst pellet has also been considered (Valdman and Hughes, 1978). For an isothermal pellet, where only internal and external mass transfer resistances were considered, the results obtained were

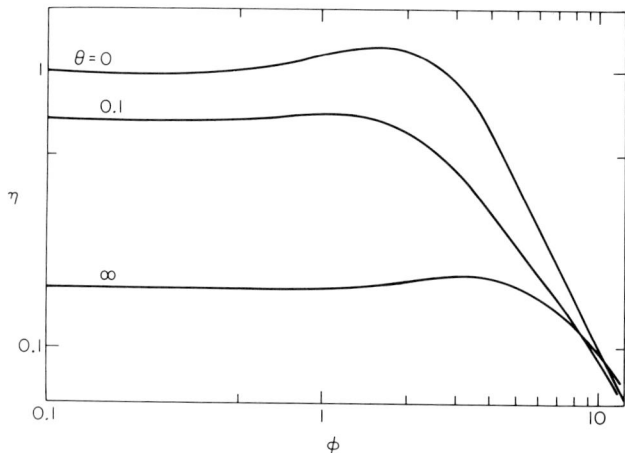

Fig. 5.9 Effect of external resistances on the reaction rate with poisoned catalysts. (Valdman, 1978)

5 Catalyst Deactivation by Poisoning

in agreement with those of Hegedus (1974). When both heat and mass transfer resistances were present, the results obtained were interpreted in the form of an effectiveness factor plot. The conditions taken for the simulation were more realistic than those used in the absence of external gradients. Thus a β value of 0.01 was employed together with values of $\gamma = 20$, $K_A C_{Ao} = 1.0$, $K_P C_{Po} = 10$, $Bi_m = 25$, and $Bi_h = 0.5$. A typical plot is shown in Fig. 5.9.

It can be seen that when no poison is present a slight maximum in the catalyst effectiveness occurs between $\phi = 0.5$ and $\phi = 3$. When the steady state poisoned condition is attained the curve still retains the same shape but with a much diminished value of the catalyst effectiveness throughout. This is because, under the poisoned conditions, diffusional mass transfer is more important, resulting in a spreading of the reaction zone and therefore distributing the production of heat throughout the pellet.

5.5 Effect of poisoning on catalyst selectivity

It is well known that diffusional resistances affect the yield of a catalytic process. In general the effect is an adverse one, in that the yield of the desired product is reduced by diffusional limitation and therefore minimization of this effect is desirable (Tamaru, 1951; Mars and Gorgels, 1964). However, it is also well known that, industrially, the selectivity of a number of processes can be enhanced by the addition of small amounts of poisons, e.g. CO and H_2S, which act as selectivity moderators.

There are two ways in which a poison may affect the selectivity of a catalyst. The first is the preferential blocking of catalyst sites in the manner described above. An alternative is that the poison reduces the deleterious effect of the diffusional limitation on the catalyst selectivity. Under the latter conditions, addition of a poison to a diffusion limited reaction may be advantageous since the increased selectivity may more than compensate for the reduced catalytic activity occasioned by the presence of poison.

A numerical analysis by Karanth and Luss (1975) has demonstrated unequivocally that a beneficial effect can occur for the consecutive purification reaction $A \rightarrow B \rightarrow C$, where A is the impurity and B is the desired reactant. This system is typical of the removal of diolefins and acetylene from olefin streams, and the usual high ratio of the reactant B concentration to that of the impurity is often the cause of severe diffusional limitation of the selectivity. It was shown by Karanth and Luss that a significant enhancement of selectivity could be obtained by the presence of poisons which compete for free sites on the surface with the reactant and thus reduce the deleterious effect of internal diffusion.

These theoretical findings have been confirmed experimentally by

Komiyama and Inoue (1970). They used a nickel catalyst for the study of the consecutive hydrogenation of acetylene, through ethylene to ethane, i.e. the system was C_2H_2 (A) $\rightarrow C_2H_4$ (B) $\rightarrow C_2H_6$ (C).

The nickel catalysts were poisoned by impregnation with $Na_2S_2O_3$ solution which effectively moderated the original high activity of the nickel catalysts. It was found that the acetylene was the main cause of catalyst deactivation and that the whole process could be described by a shell poisoning model. Discrepancies from this model could be explained by incorporating a micro-macro pore model and invoking Langmuir–Hinshelwood kinetics instead of the first order kinetics assumed originally.

Thus the effect of poisons on selectivity enhancement appears to be significant. Further work is required so that this beneficial effect of poisoning can be exploited to the full in many other processes.

References

Acres, G. J. K., Cooper, B. J., Shutt, E. and Malerbi, B. W. (1975). *Adv. Chem. Ser.* **143**, 54.
Burtonwood, P. (1979). Ph.D. Thesis, Salford University.
Butt, J. B., Downing, D. M. and Lee, J. W. (1977). *Ind. Eng. Chem. (Fund.)* **16**, 270.
Catalyst Handbook (1970). Wolfe Scientific Books, London.
Denny, P. J. and Twigg, M. V. (1980). In *Catalyst Deactivation* (ed. B. Delmon and G. F. Froment). Elsevier, Amsterdam.
Gioia, F. (1967). *Atti. Accad. Lincei* (Roma) **VIII-42**, 515.
Gioia, F. (1971). *Ind. Eng. Chem. (Fund.)* **10**, 204.
Gioia, F. and Greco, G. (1970). *Quad. Eng. Chem. Ital.* (Milano) **6**, 11.
Gioia, F., Gibilaro, L. G. and Greco, G., Jr. (1970). *Chem. Eng. J.* **1**, 9.
Hegedus, L. (1974). *Ind. Eng. Chem. (Fund.)* **13**, 190.
Hegedus, L. and McCabe, R. W. (1981). *Cat. Rev. Sci. Eng.* **23**, No. 3, 377.
Innes, W. B. (1954). *Catalysis* **1**, 245.
Jensen, J. V., Newson, E. J. and Villadsen, J. V. (1976). 3rd Int. Symp. Chem. Reaction Engng. (Heidelberg), p. 250.
Kam, E. K. T., Ramachandran, P. A. and Hughes, R. (1976). *Chem. Eng. Sci.* **31**, 247.
Karanth, N. G. and Luss, D. (1975). *Chem. Eng. Sci.* **30**, 695.
Klimisch, R. L., Summers, J. C. and Schlatter, J. C. (1975). *Adv. Chem. Ser.* **143**, 103.
Koh, H.-P. and Hughes, R. (1974). *A. I. Ch. E. J.* **20**, 395.
Komiyama, H. and Inoue, H. (1970). *J. Chem. Eng. Japan* **3**, 206.
Lundberg, W. C. (1979). *Chem. Eng. Progr.* **75** (6).
Mars, P. and Gorgels, M. J. (1964). Proc. 3rd Europ. Symp. Chem. React. Engng, p. 55.
Masamune, S. and Smith, J. M. (1966). *A. I. Ch. E. J.* **12**, 384.
Morita, S. and Inoue, T. (1965). *Int. Chem. Eng.* **5**, No. 1, 180.
Pichler, H. (1952). *Adv. Catal.* **4**, 326.
Ramachandran, P. A. (1975). *Biotechnol. Bioeng.* **17**, 211.
Shelef, M., Otto, K. and Otto, N. C. (1978). *Adv. Catal.* **27**, 311.
Tamaru, K. (1951). *Bull. Chem. Soc. Japan* **24**, 177.

Valdman, B. and Hughes, R. (1978). 6th Ibero Amer. Symp. on Catalysis (Rio de Janeiro).
Valdman, B., Ramachandran, P. A. and Hughes, R. (1976). *J. Catal.* **42**, 303.
Villadsen, J. V. and Stewart, W. (1967). *Chem. Eng. Soc.* **22**, 1483.
Weaver, E. E. (1969). SAE (Soc. Auto Eng.), Pap. No. 690016.
Wheeler, A. (1951). *Adv. Catal.* **3**, 250.

6
Catalyst Deactivation by Fouling

Catalyst deactivation by fouling, in contrast to that by poisoning, usually involves significant amounts of deposited material. Amounts up to 10–20% of the catalyst weight are frequently obtained, and apart from the resulting decrease in catalyst activity the question of pore blocking by these deposits also arises. The latter problem has already been mentioned briefly in Chapter 2.

Basically, two types of fouling may occur, due to (1) the reaction system itself and (2) deposition by an impurity in the feed stream.

The first is typified by "coke" or carbon deposition on catalysts which occurs under certain conditions when hydrocarbon streams are processed. Since the deposit originates from a cracking type reaction either in the feed stream or in the various products, it cannot be completely eliminated (but it may be minimized by varying the process conditions).

This feature distinguishes the first type of fouling from the second, since with the latter the feed may, in principle, be purified to remove the impurities. In practice this is often costly and is seldom resorted to. The classic example of this type of fouling is metal deposition from oil feedstocks which occur in petroleum processing. Usually this is manifested in the deposition of vanadium and nickel from certain crudes in hydrotreating units.

It is important to note that fouling may not occur solely as the means of catalyst deactivation. In many processes fouling, especially coking, will occur simultaneously with poisoning and also possibly with some sintering of the catalyst. An example of simultaneous coking and poisoning is that of catalysts used in steam reforming.

Because of the different characteristics inherent in the two types of fouling deposit, these will be considered separately.

6.1 Fouling by coke deposition

This type of fouling is always associated with the main reaction. Therefore it is usually not possible to eliminate the coke deposition process entirely, but

6 Catalyst Deactivation by Fouling

the process of coking can often be substantially reduced by modifying the catalyst so as to improve its selectivity. An example of this is the addition of small amounts of alkali to steam reforming catalysts. This reduces the acidity of the catalyst and tends to reduce the cracking type reactions that may occur.

Examples of reactions that produce carbonaceous deposits are extremely numerous. Virtually any process having carbon atoms in the feed or product molecules can, under appropriate conditions, give rise to deposits of coke. Naturally, molecules with large numbers of carbon atoms and/or those with aromatic or naphthenic rings tend to produce coke deposits more easily, although there is still considerable argument over the nature of the immediate "coke" precursor. Specifically, the question is whether aromatics or olefins are the immediate precursors. What is clear, however, is that both readily yield coke deposits unless care is taken in the selection of the correct processing conditions.

The time for coke levels to build up to such an extent that the catalyst activity drops to too low a level also varies considerably with the particular reaction employed. Very fast deactivation by coking occurs with fluid bed cracking catalysts. Indeed, so rapid is deactivation in this case that continuous regeneration of catalyst must be employed.

The fact that many catalysts may be regenerated (i.e. have the activity restored by burning off the coke in a mild oxidizing gas stream) is one compensating feature of deactivation by coke deposition. However, careful procedures must be adopted in this exothermic operation in order to avoid catalyst damage. This subject of regeneration is considered in detail in a subsequent chapter.

Other reactions of industrial importance in which coking is important include (a) catalytic reforming of naphthas to produce high octane number gasolines, (b) steam reforming processes, and (c) acetylene hydrogenation.

In catalytic reforming the catalyst employed is platinum supported on alumina, and although the catalyst has a long life (3 years is typical) regeneration may be necessary at intervals during this period. Unusually, with this catalyst fairly large amounts of coke can be tolerated (up to $\sim 20\%$) before catalyst activity starts to deteriorate significantly.

Steam reforming to produce syntheses gas from naphtha or methane employs a nickel catalyst. This is subject to coke deposition and may also be deactivated by very small amounts of sulphur remaining in the feed stream. The similar methanation reaction, which operates at lower temperatures but with a similar catalyst, may also be deactivated by coke deposition.

Acetylene hydrogenation, as would be expected from the nature of the process, is also subject to extensive coking.

6.1.1 Coke formation processes

A number of investigators have examined the nature of the coke deposit on catalysts. Microscopic examination by Haldeman and Botty (1959) has established that the coke consists of filmy aggregates of size less than 10 nm. This discrete nature of the coke deposits has been confirmed by many other investigations. X-ray studies have also established that approximately 50% of the coke deposits are in the form of pseudo-graphitic structures with the residue probably existing as unorganized aromatic systems and of aliphatic and alicyclic appendages to polynuclear aromatic systems. It is generally agreed that the molecular formula of coke deposits varies from $C_1H_{0.4}$ to C_1H_1. Thus the coke corresponds to a hydrocarbon in composition but with only a small amount of hydrogen.

For coking to low levels with fresh catalyst, Haldeman and Botty (1959) found no significant change in surface area or pore volume. It seems that under these conditions for the particular catalyst used the pores are not sealed off by coke particles. However, Ramser and Hill (1958) found a reduction in surface area of 20% and of pore volume of 22% when an aged catalyst was coked to 2.2 weight %. It appears that in this case the aged catalyst had sintered and the pores had become sealed off by deposits of coke near the boundaries of the active region. In general, unless large amounts of coke are formed, the physical properties of the catalyst pellets are not significantly changed and therefore transport restrictions are also generally not important.

The precursor to the coke deposits is still subject to some dispute. For some time it was believed that, because of the aromatic condensed ring form of many coke aggregates, aromatics were the immediate precursors. More recently, Wojciechowski *et al.* (1974) challenged this and proposed that the immediate precursors are olefins. Their arguments were based on an analysis of their experimental curves obtained by cracking typical oil feedstocks. They showed that coke was not a primary product of the reaction but was a product of secondary reactions arising from the primary products. Since the only primary products that were qualitatively different from the feed in the experiments were olefins, it was inferred that coke formation was attributable to these.

There seems little doubt that it is difficult to ascribe coke formation to any single precursor. The nature of the feedstocks used in experiments has varied considerably and it is probable that many mechanistic proposals may be put forward to explain the development of coking under specified conditions. It should be noted that all hydrocarbons, from paraffins to aromatics, will yield coke under appropriate experimental conditions.

6.1.2 Experimental results from laboratory coking studies

The work of Voorhies (1945) has been mentioned previously. His correlation was based on experimental measurements made in laboratory and semi-pilot scale reactors. The small temperature coefficient he observed was explained by a diffusional control reaction, and since he found that the rate of carbonization decreased as the percentage of carbon on the catalyst increased, he postulated that the coke itself was the diffusion barrier. The independence of coke formation on feed rate was explained on the basis of different products within the catalyst bed. Thus, the two main types of product would be (a) gasoline and liquid petroleum gas (LPG) which have less tendency to carbonize than the original feed and (b) cycle gas oil which has more tendency to coke formation than the original feed. These two effects would tend to balance one another and the extent of coke formation on the catalyst would not differ at any level in the catalyst bed at any given time. This implies that coke deposition on the catalyst would be independent of feed rate, within limits. Eberly and coworkers (1966) showed that there was some dependence of coking on space velocity, as might be expected.

Neither of these groups of workers examined the distribution of coke within the catalyst bed. In a theoretical analysis, Froment and Bischoff (1961) examined this problem for a fixed bed reactor. By assuming that coking would occur in a reaction either in parallel or consecutive to the main reaction, it was shown that a profile of deposited coke would exist within the reactor. This coke profile would be descending in the case of a parallel mechanism but ascending for a consecutive (or series) mechanism.

The parallel and consecutive reactions for coking can be written as follows:

$$\left.\begin{array}{l} A \to B \\ A \to C \text{ (coke)} \end{array}\right\} \text{Parallel fouling}$$

$$A \to B \to C \text{ (coke)} \quad \text{Series fouling}$$

Thus, parallel fouling gives large coke deposits when the reactant concentration is high, since the reactant is the coke precursor. Therefore, when coking occurs by a parallel mechanism, the greatest deposition of coke would be expected at the inlet of the reactor as was predicted by Froment and Bischoff (1961). Conversely, larger coke deposits are formed in series fouling, when the product B has a high concentration since this is the immediate precursor of the coke in this case. In normal operation the product concentration increases with distance along the reactor, and therefore the coke distribution should follow a similar pattern.

A distribution of coke deposits along the reactor was also observed by

Noda et al. (1974) in a study of isopentane isomerization. In this case the coke deposit increased with axial distance from the reactor inlet, suggesting that coking was occurring by a series mechanism.

To determine whether a corresponding distribution existed within single catalyst pellets, Richardson (1972) used a regeneration technique to determine the carbon concentration in different layers of the single pellet. For this study the supported cobalt molybdate catalyst was coked using a coal derived liquid. The catalyst spheres were then regenerated under diffusion controlled conditions in an oxygen atmosphere. After a short burning time the regeneration was stopped, the sphere cut open, and the radius of the well defined shell of carbon measured. The carbon dioxide liberated during the regeneration was collected and measured. By repeated application of this technique for successively longer regenerations and for a large number of samples, a total carbon vs. radius curve was constructed from which the carbon concentration profile was obtained. The resultant profiles showed that carbon deposition was greatest in the outer layers of the pellet, suggesting that coking occurred by a parallel mechanism in this case.

The most explicit experimental demonstration of the differing effects of fouling by parallel and series mechanisms was given by Murakami and coworkers (1968). They considered two processes, typical respectively of parallel and of series coking. For parallel coking the reaction considered was the disproportionation of toluene which proceeds by the following scheme:

$$\text{Toluene (A)} \rightarrow \text{Xylene (B)} + \text{Benzene}$$

$$\text{Toluene (A)} \rightarrow \text{Coke (C)} + \text{Benzene}$$

where the xylene (B) represents the derived product and C is the coke deposit.

For series, or consecutive, fouling a representative reaction is the dehydrogenation of primary alcohols. In this type of reaction the intermediate aldehyde product may itself be dehydrogenated and condensed on the active sites of the catalyst, being finally converted into coke. This final coke product does not desorb from the active sites and therefore reduces the activity of the catalyst. Thus, for this series fouling the process may be depicted as:

$$\text{Alcohol (A)} \rightarrow \text{Aldehyde (B)} \rightarrow \text{Coke (C)}$$

For the disproportionation of toluene an alumina–boria catalyst containing 10% of boria was used by Murakami et al. (1968). The results obtained by cross-sectioning the pellets at two different temperatures are shown in Fig. 6.1(a). In a parallel deactivation process the reactant A will have the higher concentration in the outer part of the catalyst pellet. Furthermore, the concentration gradient of A in these outer layers of the catalyst pellet will increase with increasing values of the Thiele modulus, ϕ. When ϕ becomes

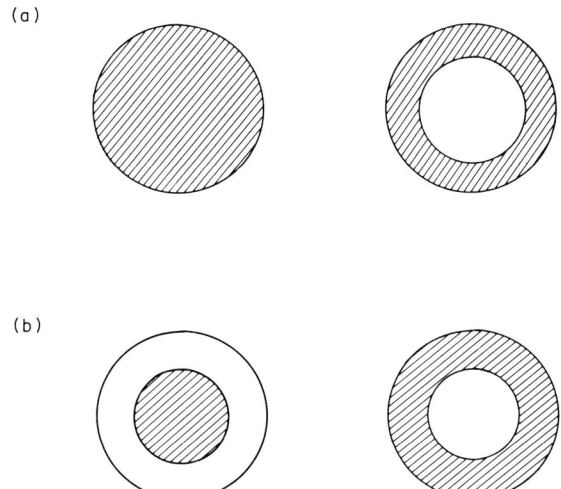

Fig. 6.1 (a) Cross section of catalyst pellet after 10 min reaction (disproportionation of toluene): left 440°C; right 530°C.
(b) Cross section of catalyst pellet after dehydrogenation of n-butyl alcohol: left 400°C after 50 min; right 480°C after 30 min. The shaded parts represent coke deposited. (Murakami *et al.*, 1968)

very large the coked layer may be distinct from the uncoked layer, while at very low values of ϕ, with no significant diffusional resistance, coke deposition may be uniform throughout the pellet. Figure 6.1(a) illustrates these two extremes. For this example the increase in Thiele modulus is obtained by increasing the reaction temperature. Thus, at 440°C, when the Thiele modulus would be expected to be small, the pellet appeared to be coked throughout, but at 550°C, where a larger diffusional restriction operates, an outer coked zone and a separate inner coke-free zone are clearly demonstrated.

For a series fouling process the coke is formed from the intermediate B. For moderate values of the Thiele modulus the concentration of B is greatest at the centre of the pellet, and therefore coke should form there initially and spread out towards the pellet surface as reaction proceeds. At high values of the Thiele modulus the diffusional restriction reverses this process, and both reactant and product are present in the outer layers of the pellet. This behaviour is shown in Fig. 6.1(b) where, as for parallel fouling, differences in ϕ were achieved by varying the reaction temperature. At 400°C the value of ϕ is small and the central coked area is quite distinct, but at 480°C the diffusional resistance is controlling and the coke deposit is now only observed in the outer layers of the pellet.

6.1.3 General approaches to developing a theory for deactivation by coking

Because of its industrial importance a large number of attempts have been made to develop predictive methods for catalyst coking. There has been a certain amount of controversy in this area, which centres on whether time or the concentration of coke is the key parameter affecting catalyst deactivation by coking. It might seem that these are much the same and either could be used to represent the extent of deactivation, but equivalence of the two measures would only be true if the coke concentration were linear in time. In most cases it is not.

Most of the early work was necessarily empirical, and since the key question to which a plant operator wishes an answer is "How long do I leave the catalyst on stream?", time was the natural parameter to use. Hence, the results of Voorhies (1945) were expressed using time as parameter, as already noted, and a number of other investigators have also used time as parameter in their correlations for activity decrease.

Wojciechowski has summarized his extensive results on cracking of oils over various catalysts in a review paper (1974). This expounds his "time on stream" theory of catalyst decay which, although usually applied to coking, may also be employed for suitable poisoning processes. The time on stream theory specifies that for deposition reactions the amount of deposit formed is governed by the time of exposure. This will be justified if (for poisoning reactions) the feed contains a constant amount of poison or if the feed and products are equally poisonous towards the catalyst. In the time on stream theory the activity is defined in terms of the active sites concentration. A further assumption is that the activity of all sites is the same. Wojciechowski and coworkers (Pachovsky et al., 1973) assumed that a general expression for the rate of catalyst decay could be written:

$$-\frac{d[S]}{dt} = k_d[S]^m C_d \qquad (6.1)$$

where S represents the active sites and C_d is the concentration of deactivating species.

If the fraction of sites θ available at any time ($\theta = [S]/[S_o]$) is substituted into (6.1) we obtain:

$$-\frac{d\theta}{dt} = k'_d \theta^m \qquad (6.2)$$

where $k'_d = k_d[S_o]^{m-1} C_d$, and C_d is assumed constant.

Equation (6.2) may then be integrated for different values of m. For $m = 1$

this leads to:
$$\theta = \exp(-k'_d t) \tag{6.3}$$
and for $m \neq 1$, equation (6.2) gives:
$$\theta = [1 + (m-1)k'_d t]^{-[1/(m-1)]} \tag{6.4}$$

The exponential form (6.3) has been used with some success by Weekman (1968) to correlate activity in cracking reactions. The more general hyperbolic form may be rewritten more conveniently as:
$$\theta = (1 + Gt)^{-M} \tag{6.5}$$
where $G = (m-1)k'_d$ and $M = 1/(m-1)$ are measures of the deactivation rate constant and the order of the deactivation reaction respectively.

Substitution for [S] into the rate expression for a reaction A → B of order n:
$$-r_A = k[S]^n (C_A)^a \tag{6.6}$$
gives
$$-r_A = k[S_o]^n (1 + Gt)^{-nM} (C_A)^a$$
or
$$-r_A = k^o (1 + Gt)^{-N} (C_A)^a \tag{6.7}$$
with $N = -nM$.

Results have been obtained for different orders of the main reaction and for different values of m. Some results from Bischoff (1976) are given in Fig. 6.2 for m values <1, 1, and >1 for a first order main reaction and for a constant catalyst/oil ratio in a plug flow reactor. The different shaped curves for the three classes of deactivation are apparent. A number of successful correlations for various reactions and different catalysts have been reported by Wojciechowski and coworkers and the method appears to be useful. However, it should be borne in mind that the method does involve four adjustable parameters in most applications.

Froment (1976) and Froment and Bischoff (1961) have argued consistently that correlations between coke content of the catalyst and the process time lack generality. They prefer to relate the coke concentration to process variables such as the partial pressures of reactants and products, space time, etc., and point out that time is not a true variable in these systems. Use of coke concentration as a variable enables the equations to be written in Langmuir–Hinshelwood form in the same way as the equations that are applicable for the main reaction in many instances.

Froment and Bischoff (1961) also introduced a deactivation function Φ, to

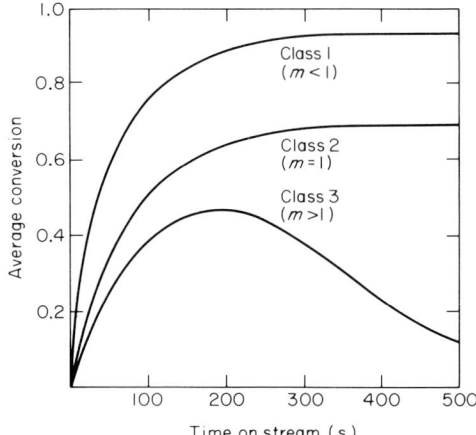

Fig. 6.2 Average conversion *vs.* catalyst time-on-stream, illustrating the three classes of ageing catalysts.

account for the influence of coking on the main reaction. This function is an empirical multiplying factor which is used to relate the non-deactivated reaction coefficients to that when deactivation occurs:

$$k = k_i^o \Phi(C_c) \qquad (6.8)$$

where k_i^o is the value of the rate coefficient of the main reaction when the coke content is zero and C_c is the coke content. The precise form of the function $\Phi(C_c)$ depends upon conditions. One possibility is a linear form:

$$\Phi = 1 - \alpha C_c \qquad (6.9)$$

where α is a constant. Such a relation was observed by Ozawa and Bischoff (1968) in their study of the cracking of ethylene over a silica–alumina catalyst and by Weekman and Nace (1970) in their study of oil cracking. In their earlier work Froment and Bischoff (1961) had proposed two alternative forms for Φ:

$$\Phi = \exp(-\alpha C_c) \qquad (6.10)$$

$$\Phi = \frac{1}{1 + \alpha C_c} \qquad (6.11)$$

de Pauw and Froment (1975) studied the isomerization of pentane over a platinum reforming catalyst in the presence of hydrogen. When operated with large amounts of hydrogen no deactivation occurs, so the kinetics of the main reaction may be determined readily. However, at low hydrogen/

pentane ratios severe coking occurs and the system lends itself to a study of deactivation by coking. de Pauw and Bischoff found that an exponential form of the deactivation best fitted their results for this system. Even so, the determination of the best fit rate expression involves the use of a number of parameters in the Langmuir–Hinshelwood rate expression.

An alternative approach was adopted by Kam *et al.* (1975, 1977a,b, 1979) in which the main reaction is written in terms of the availability of active sites and the site concentration is itself a decaying function with time which may be generalized to a parallel mechanism or a series mechanism or a combination of both. This approach is discussed in detail in the following section which considers modelling processes for coking, so it will not be considered further here. One consequence of this method is that the rate constants for the fouling process must be estimated in some way.

To summarize, there are a number of theoretical treatments of deactivation by coking. All, however, require parameter estimation in order to obtain predictive results. Although coke deposition may be the "true" variable in the deactivation process, this has to be linked to time by some form of deactivation to obtain some indication of how long a catalyst may remain on stream before replacement or regeneration is necessary. There is clearly a need for much more experimental data especially of the reaction type investigated by de Pauw and Froment where the non-deactivated reaction kinetics may be determined separately from those when deactivation occurs. Future progress is likely to hinge on experimental work of this kind.

6.1.4 Modelling of coking in single catalyst pellets

Despite the lack of agreement in general theories of catalyst decay due to coking, there has been considerable progress in the modelling of reactions in single catalyst pellets subject to coking.

One of the most important analyses of fouling in single catalyst pellets was that of Masamune and Smith (1966). This has already been mentioned briefly in Chapter 5 when poisoning was discussed, since these authors also considered this aspect of catalyst deactivation in their analysis. Isothermal conditions were assumed throughout the pellet and gas film, and Masamune and Smith start with the usual division of fouling into parallel and series processes:

$$A(g) \rightarrow B(g) \quad \text{Main reaction}$$
$$A(g) \rightarrow C(s) \quad \text{Parallel fouling}$$
$$B(g) \rightarrow C(s) \quad \text{Series fouling}$$

Here C(s) is the solid (coke in this case) deposited on the catalyst. The effect of deactivation on the main reaction is assumed to be linear in form. If

q_o is the concentration of C on the surface when deactivation is complete, the deactivation function is:

$$\Omega = 1 - \frac{q}{q_o} = 1 - \psi \qquad (6.12)$$

If the main reaction is assumed to be first-order, the rates of the main reaction and the parallel and series fouling reactions are given by:

$$r = k_A C_A (1 - \psi) \qquad (6.13)$$

Series fouling: $\quad \dfrac{dq}{dt} = k_{Bf} C_B (1 - \psi) \qquad (6.14)$

Parallel fouling: $\quad \dfrac{dq}{dt} = k_{Af} C_A (1 - \psi) \qquad (6.15)$

These rate expressions are coupled with the conservation equations for A and B within the pellet. These should be written with a transient term for A and B, but since the time for the accumulation of mass within the pellet is negligible compared with the time required for the catalyst activity to change significantly, the pseudo-steady state assumption may be invoked for species A and B and the transient terms can then be neglected. The consequent steady state conservation equations from A and B then become:

$$D_{eA} \nabla^2 C_A - \rho k_A C_A (1 - \psi) = 0 \qquad (6.16)$$

$$D_{eB} \nabla^2 C_B + \rho k_A C_A (1 - \psi) = 0 \qquad (6.17)$$

Film resistances are assumed absent, so the usual boundary conditions apply together with the appropriate initial condition that the activity is unity at zero time:

$$\psi = 0; \quad t = 0; \quad r_o \geqslant r \geqslant 0 \qquad (6.18)$$

$$C_A = C_{Ao}; \quad C_B = C_{Bo}; \quad t \geqslant 0; \quad r = r_o \qquad (6.19)$$

$$\frac{dC_A}{dr} = \frac{dC_B}{dr} = 0; \quad t \geqslant 0; \quad r = 0 \qquad (6.20)$$

Constant diffusivity is assumed for A and B, i.e. deposition of coke has no effect on transport within the pores.

The effectiveness factor, η, defined as the rate of the overall main reaction for the pellet divided by the rate for conditions at the surface of the pellet, and at $t = 0$, is expressed as:

$$\eta = \frac{4\pi r_o^2 D_{eA} \left(\dfrac{dC_A}{dr}\right)\Big|_{r=r_o}}{\frac{4}{3}\pi r_o^3 \rho k_A C_{Ao}} = \frac{3}{\phi^2} \frac{dC_A^*}{d\delta}\bigg|_{\delta=1} \qquad (6.21)$$

6 Catalyst Deactivation by Fouling

where C_A^* and δ are the dimensionless concentration of A and the radius of the pellet respectively.

Numerical solutions for the concentration profiles of A and B and the residual activity profiles are given in Fig. 6.3(a) for series fouling and Fig. 6.3(b) for parallel fouling. Both sets of profiles are for intermediate values of the Thiele modulus ($\phi = 5$). Other parameters included in the analysis include the dimensionless initial concentration ratio, C_{Bo}/C_{Ao}, and the diffusivity ratio, D_{eA}/D_{eB}. Both of these quantities were set equal to unity for series fouling.

Figure 6.3(a) shows that, for series fouling, since the product B must

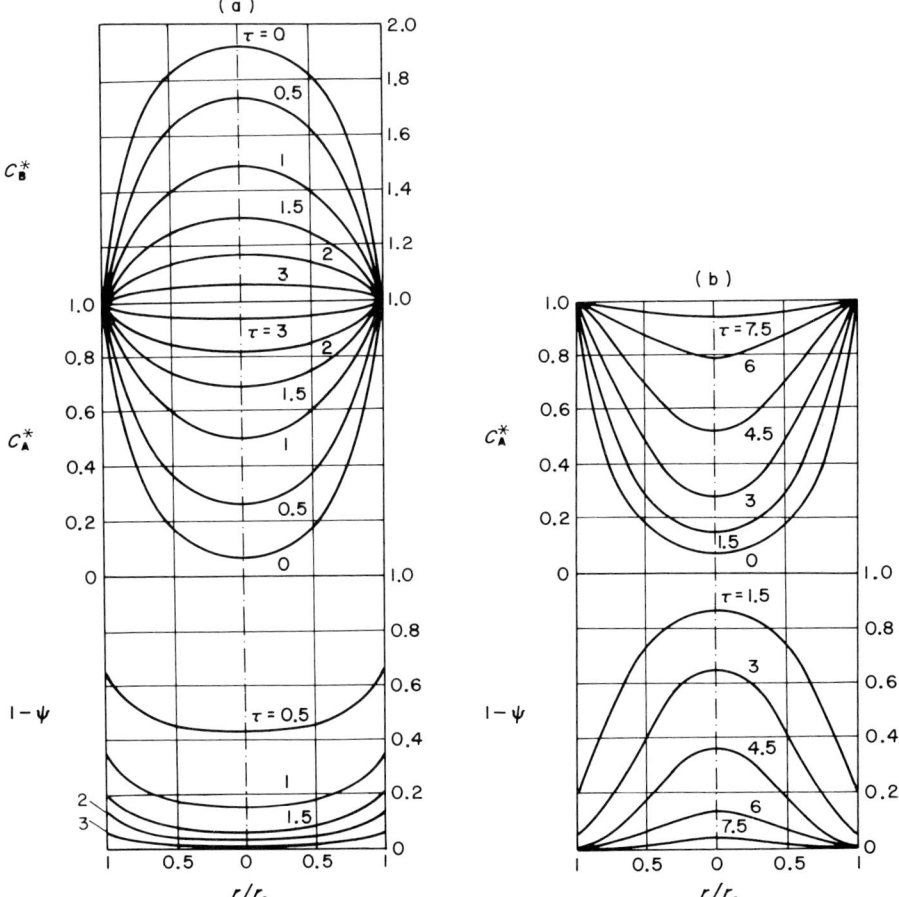

Fig. 6.3 Profiles for self-fouling: (a) series mechanism, $\phi = 5$; (b) parallel mechanism, $\phi = 5$. (Masamune and Smith, 1966)

diffuse out of the pellet its concentration decreases from a maximum value at the centre to the surface. This high concentration of B at the pellet centre means, in turn, that the central part of the pellet is more seriously fouled than the outer layers. This is reflected in the residual activity profile $(1 - \psi)$ which has a minimum at the centre of the pellet. On all these curves the numbers refer to different fouling times.

The corresponding profiles for parallel fouling are shown in Fig. 6.3(b). Parallel fouling is simpler than series fouling since there is no interaction between A and B. The profile of A has, as expected, a maximum at the surface. The corresponding activity profiles will therefore show a minimum at the surface and a maximum at the pellet centre.

To compare the effect of varying the Thiele modulus the effectiveness factor was plotted as a function of the dimensionless time, τ, with Thiele modulus as parameter. Figure 6.4(a) shows the plots for series fouling at a value of the ratio C_{Bo}/C_{Ao} equal to 10. The dotted curve represents the results for negligible diffusion resistance, i.e. $\phi = 0$. The extent of deactivation is seen to increase markedly with ϕ and is independent of process time, so the preferred catalyst is one with the least diffusion resistance. However, the large value of C_{Bo}/C_{Ao} used in this result would only be realized in practice for large values of a recycle ratio of the product stream.

The corresponding plot for parallel fouling is shown in Fig. 6.4(b) and again the dotted curve represents a Thiele modulus of zero. In contrast to the results for series fouling, there is now a crossover of the curves for different ϕ values. Thus for fresh catalysts there is a continuous decrease in effectiveness factor as the diffusion resistance increases, but as the process time is increased the effectiveness factor first increases and then decreases with ϕ. It seems probable therefore that fresh catalysts with only a small diffusional resistance are fouled more rapidly than those with a large diffusional resistance.

In a later paper, Smith and coworkers (Sagara *et al.*, 1967) extended the analysis to non-isothermal systems but still assumed that film resistances to both mass and heat transport were absent. The same rate expressions, equations (6.13)–(6.15), and the same conservation equations (6.16) and (6.17), were used as in the earlier analysis, but an additional heat balance has now to be employed:

$$K_e \nabla^2 T + (-\Delta H)\rho k_A(1 - \psi)C_A = 0 \qquad (6.22)$$

However, when the equations are made dimensionless new parameters arise because of the non-isothermal nature of the system. These parameters are those due to the heat of reaction (β) and to the activation energies of the main (γ) and fouling reactions (γ_f).

The system of equations involving concentrations of two species and

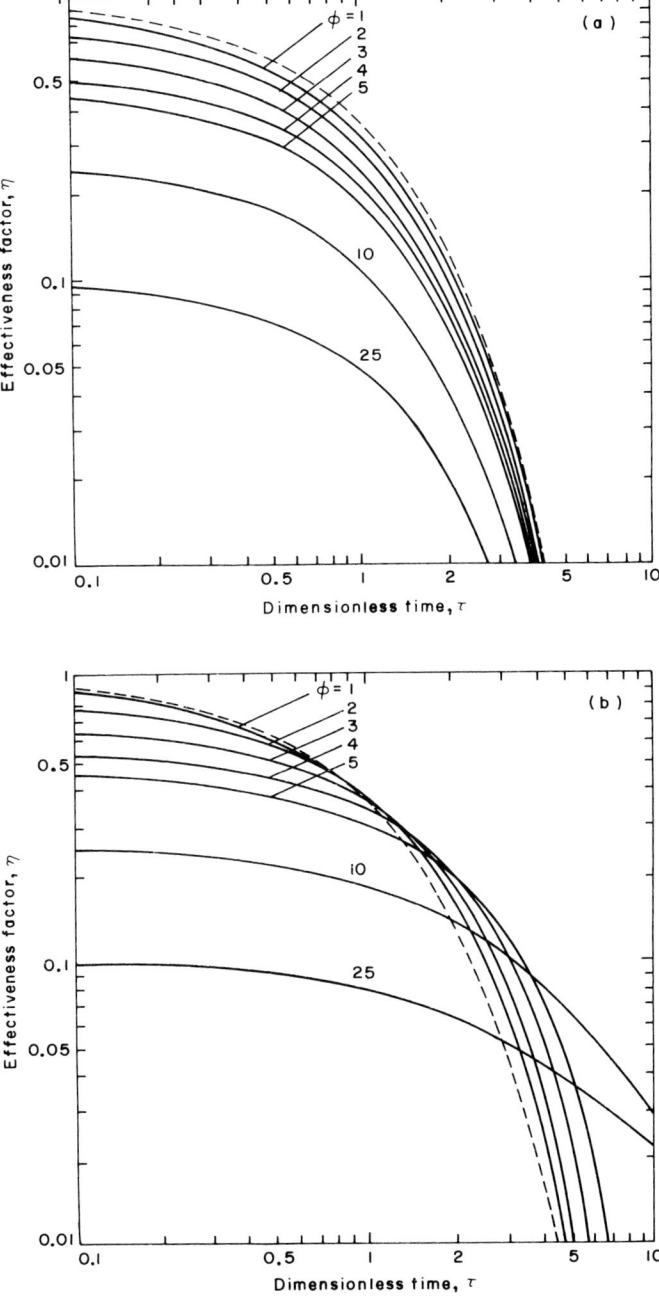

Fig. 6.4 Effectiveness factor for (a) series fouling, $C_{Bo}/C_{Ao} = 10$, and (b) parallel fouling. (Masamune and Smith, 1966)

temperature may be simplified by use of the Prater relation which relates concentration to temperature in a catalyst pellet under steady-state conditions. Thus, either temperature or concentration may be eliminated with a consequent easier solution.

Profiles of coke concentration and dimensionless temperature and reactant concentration are given in Fig. 6.5(a) for parallel fouling. Other values selected were $\beta = +0.1$, i.e. exothermic reaction, and $\gamma = 20$ with $\gamma_f = 30$. The latter values correspond to activation energies of about 80 and 120 kJ mol^{-1} at a surface temperature of 500 K. A moderate value of the Thiele modulus of 5 was also adopted in these simulations, and the dimensionless process time is the parameter on all the illustrated profiles. As can be seen, the interaction of temperature with diffusional resistance leads to complex radial profiles of the deposited material, ψ. The pronounced drop in reactant concentration from the surface to the centre of the pellet at low process times reflects the moderate value of ϕ chosen. The analogous increase in temperature is also shown. The interaction of these two effects on the fouling reaction leads to the unusual profiles in ψ in Fig. 6.5(a). Near the outer surface of the pellet the temperature increase more than offsets the decrease in concentration, so the rate of fouling increases ($\gamma_f > \gamma$). Further inside the pellet the concentration effect is dominant. Hence, ψ goes through a maximum and then decreases towards the centre of the pellet. With increase in time the maximum becomes less pronounced and disappears at a process time of 3.

For series fouling, the product B is the coke precursor and its concentration is greatest at the centre of the pellet. Therefore, for an exothermic reaction the effects of temperature and concentration now act together and cause increased fouling towards the centre of the pellet. The profiles are shown in Fig. 6.5(b) for an initial concentration ratio of 1.0 and show that both the concentration of B and deposition go through a maximum while A goes through a minimum.

The isothermal analysis of Masamune and Smith was extended by Murakami and coworkers (1968) in two important respects. First, a full transient analysis was adopted with no pseudo-steady assumption, so the results obtained are valid for fast fouling conditions at short contact times. Secondly, external film resistances were included in the formulation. One of the most interesting aspects of these authors' results was the theoretical confirmation of the experimental results for series fouling shown in Fig. 6.1(b) where, at high values of the Thiele modulus, the strong diffusional effect overcomes the normal feature of a maximum of deposit at the centre of the pellet and gives a resultant coke profile reversal. The numerical results are shown in Fig. 6.6 where the left-hand diagram is for a low ϕ value equal to 2 while that for the right-hand diagram is for $\phi = 20$. A comparison with Fig.

6 Catalyst Deactivation by Fouling

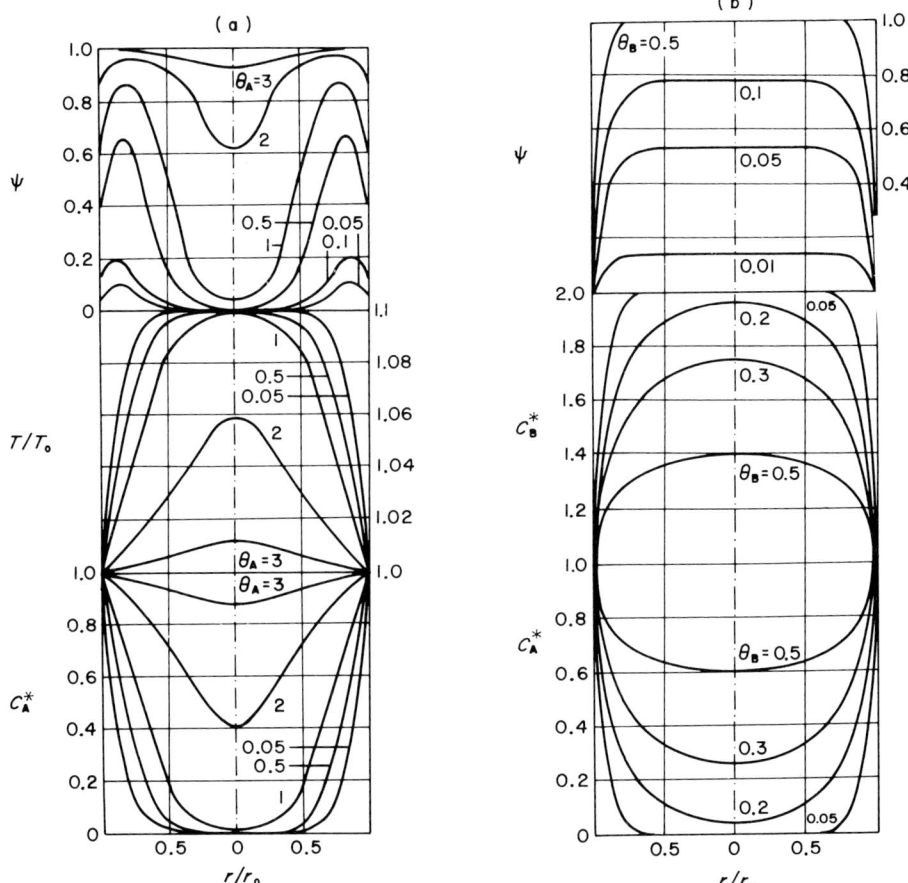

Fig. 6.5 Radial profiles, $\gamma = 2$, $\gamma_f = 30$, $\beta = 0.1$, for (a) parallel fouling, $\phi = 5$, and (b) series fouling, $\phi = 10$, $C_{Bo}/C_{Ao} = 1.0$. (Sagara et al., 1967)

6.1(b) shows that good agreement between theory and experiment was obtained.

The effect of external film resistance in the work of Murakami et al. (1968) was found not to be too significant. A variation of the modified Sherwood number Sh* (Sh* $= k_e R/D_e$) from 10 to 100 gave only slight differences in effectiveness factor under fouling conditions. This result has since been confirmed by other workers.

An extensive study of the modelling of coking processes on catalyst pellets has been made by Kam and Hughes and coworkers. Starting initially with an isothermal pellet, the analysis was successively extended to include non-

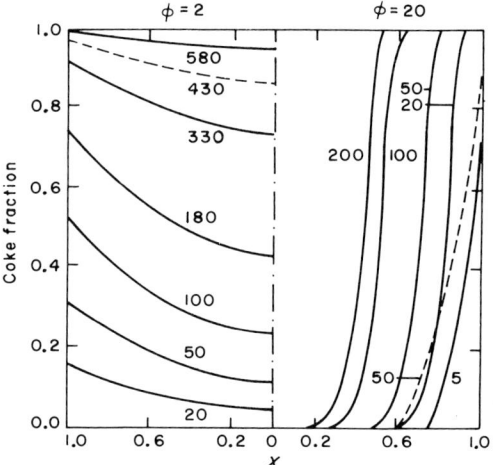

Fig. 6.6 Coke profiles in a pellet for various values of dimensionless time. The dashed lines represent concentration of A. Parallel reaction; $s = \rho = 1$, $N_B = 20$, $q = 1/500$. (Murakami et al., 1968)

isothermal behaviour, non-isothermal behaviour with film effects, transient analysis, and finally a full Langmuir–Hinshelwood analysis of fouling.

The basis of the modelling was to assume, as most other workers have done, that fouling could occur by a reaction in parallel or series to the main reaction. Since many cracking reactions occur by a complex series of reaction pathways involving both series and parallel reactions, the possibility of simultaneous fouling by a parallel/series process was also included. The rate of the main reaction was assumed to be a function of the number of active sites available for reaction. These active sites could be removed by either series or parallel fouling processes.

For the isothermal analysis of a single pellet (Kam et al., 1975), external mass transfer was also neglected. Langmuir–Hinshelwood kinetics were assumed for the main reaction, A → B + C, while general order kinetics are assumed for the parallel and series deactivation processes, A → Coke and B → Coke, respectively.

The mass balance for the reactant A and the product B in the catalyst pellet for the irreversible main reaction can be formulated as:

$$D_{eA}\nabla^2 C_A = \varepsilon \frac{\partial C_A}{\partial t} + r_A(C_A)S \quad (6.23)$$

$$D_{eB}\nabla^2 C_B = \varepsilon \frac{\partial C_B}{\partial t} - r_A(C_A)S \quad (6.24)$$

The fouling reaction results in the occupation by a foreign deposit of the active sites available for the reaction. For the purpose of modelling, a suitable rate equation has to be postulated for the fouling reaction which is generally a function of concentration of the species involved in the deactivation reaction, and also the amount of active sites present at any given time. Thus, without loss of generality, the decrease in the fraction of the active sites can be expressed by a relation of the following form:

$$-\frac{dS}{dt} = k_{f1} f_1(C_A) g_1(S) + k_{f2} f_2(C_B) g_2(S) \tag{6.25}$$

where k_{f1} and k_{f2} are the rate constants for parallel and series fouling respectively.

The boundary conditions at any time are:

$$\text{At } r = 0 \quad \frac{\partial C_A}{\partial r} = \frac{\partial C_B}{\partial r} = 0 \tag{6.26}$$

$$\text{At } r = R \quad C_A = C_{Ao}; \quad C_B = C_{Bo} \tag{6.27}$$

where R is the radius of the pellet and the initial condition at any r is:

$$\text{At } t = 0 \quad S = 1 \tag{6.28}$$

Solution of equations (6.23)–(6.25) simultaneously gives the concentration and active site profiles in the pellet as a function of time. Knowing the concentration distribution inside the pellet, the rate of reaction and hence the effectiveness factor of the catalyst at any time can be evaluated.

An assumption implicitly made in equation (6.23) is that the fouling reaction rate is very small in comparison with the main reaction, and hence the amount of A consumed in the deactivation reaction does not appear in the mass balance equation for A. Under these conditions a pseudo-steady state assumption can be made for the main reaction, and the transient terms in equations (6.23) and (6.24) can be neglected.

The equations can be made dimensionless by the introduction of the following variables:

$$a = \frac{C_A}{C_{Ao}}; \quad b = \frac{C_B}{C_{Bo}}; \quad \delta = \frac{r}{R}$$

$$\tau = k_{f1} C_{Ao} t \quad \text{(parallel fouling)}$$

$$\tau = k_{f2} C_{Ao} t \quad \text{(series fouling)}$$

Thus equation (6.23) becomes:

$$\nabla^2 a = \frac{R^2}{D_{eA} C_{Ao}} r_A(a) S \tag{6.29}$$

where the Laplacian operator is now with respect to S. Similarly equation (6.24) becomes:

$$\nabla^2 b = \frac{D_{eA}}{D_{eB}} \frac{R^2}{D_{eA} C_{Ao}} r_A(b) S \tag{6.30}$$

and equation (6.25) becomes:

$$-\frac{dS}{d\tau} = f_1(a) g_1(S) + \frac{k_{f1}}{k_{f2}} f_2(b) g_2(S) \tag{6.31}$$

The dimensionless boundary conditions are:

$$\text{At } \delta = 0 \quad \frac{da}{d\delta} = \frac{db}{d\delta} = 0 \tag{6.32}$$

$$\text{At } \delta = 1 \quad \begin{aligned} a &= a_o = 1 \\ b &= b_o \, (= C_{Bo}/C_{Ao}) \end{aligned} \tag{6.33}$$
$$\tag{6.34}$$

and the dimensionless initial condition is:

$$\text{At } \tau = 0 \quad S = 1 \tag{6.35}$$

Equations (6.29) and (6.30) can be combined to eliminate the rate terms, and the concentration b can be expressed in terms of the concentration a:

$$b = \frac{D_{eA}}{D_{eB}} (1 - a) + b_o \tag{6.36}$$

The problem then consists of solving equations (6.29) and (6.31) sequentially to evaluate the concentration distribution of A and the active sites profile inside the pellet. This was done using the method of orthogonal collocation (Villadsen and Stewart, 1967). With the concentration profile and active sites distribution known, the effectiveness factor can be found from:

$$\eta(\tau) = \frac{\alpha \int_0^1 \delta^{(\alpha-1)} r_A(a) S \, d\delta}{r_A(a)|_{\delta=1}} \tag{6.37}$$

where α is the geometrical factor, dependent on the shape of the pellet ($=1, 2,$ or 3 for slabs, cylinders, or spheres respectively).

The main reaction was assumed to be of the Langmuir–Hinshelwood form (Beranek, 1972) typical of the dehydration of alcohols:

$$r_A(C_A) = \frac{k C_A^{\frac{1}{2}}}{1 + K_A C_A^{\frac{1}{2}} K_W C_W} \tag{6.38}$$

where the subscript A indicates alcohol and W the product water.

Results for this simple isothermal model are shown in Fig. 6.7 where the pellet effectiveness factor is plotted against the Thiele modulus for the main reaction. The variation of the catalyst activity, S, with time at various radial positions in the pellet is shown in Fig. 6.8 for three Thiele moduli. The extent of fouling and the resulting distribution of the active sites at any given time depends on the Thiele modulus. For small values of the Thiele modulus, the reaction of A takes place in the entire volume of the pellet and the effectiveness factor, η, tends to unity at zero time because of the lack of diffusional restrictions. For parallel fouling, the fouling is directly related to the concentration of reactant A, and consequently the fouling is also uniform throughout the pellet as evidenced by the virtual horizontal lines in Fig. 6.8(a). At large values of the Thiele modulus, the reaction of A occurs over a small region near the surface of the pellet. This results in the deactivation being confined to this same small region near the surface whilst the catalyst in the interior of the pellet still retains the original activity. This situation, illustrated in Fig. 6.8(c), corresponds to the pore mouth deactivation of Wheeler (1955). With increase of time the deactivated zone moves into the interior of the pellet. For intermediate values of the Thiele modulus, fouling occurs within the centre of the pellet but is greater near the surface; Fig. 6.8(b). Analogous behaviour is observed when the effectiveness factor is plotted against the Thiele modulus as in Fig. 6.7. For small values of the Thiele modulus (<2), the value of the effectiveness factor decreases with

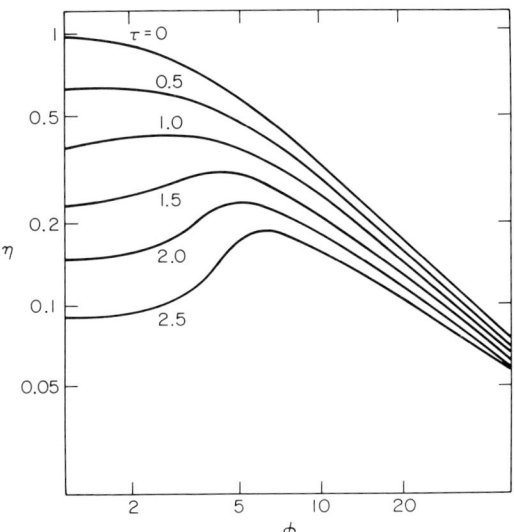

Fig. 6.7 Plot of η vs. ϕ for parallel fouling. (Kam et al., 1975)

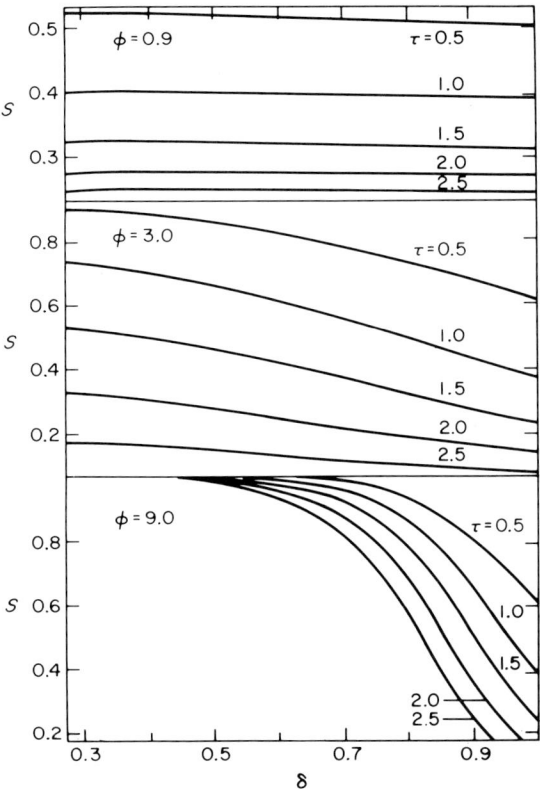

Fig. 6.8 Active sites distribution for parallel fouling. (Kam et al., 1975)

time. At higher values of the Thiele modulus the reduction in the effectiveness factor may be much less pronounced with increase in fouling time. This arises because, under these conditions, fouling approaches pore mouth deactivation behaviour which results in the bulk of the catalyst in the pellet interior having a high residual activity. This point may be demonstrated from Fig. 6.7 by considering values of ϕ equal to 1 and 5. When the pore diffusion resistance is very small ($\phi = 1$), the effectiveness factor values for time $\tau = 0$ and $\tau = 2$ are 1.0 and 0.15 respectively. When the Thiele modulus in the catalyst pellet has a value of 5, the initial effectiveness factor is 0.58 because of pore diffusion restrictions. However, at a time $\tau = 2$, the effectiveness factor at this same Thiele modulus has decreased to only 0.25, which is greater than that for the catalyst pellet which had little or no pore diffusion resistance initially ($\phi = 1$). The activity of the catalyst remaining after a given process time may be obtained from the area under the curve of S

6 Catalyst Deactivation by Fouling

against δ for the appropriate value of τ chosen. For a fouling time of $\tau = 2$, the areas are 0.4 for the ϕ value of 1 and 0.85 for the ϕ value of 5. This explains the higher value of the effectiveness factor at $\phi = 5$ compared with $\phi = 1$ under conditions of catalyst fouling. The extreme case is for values of the Thiele modulus in excess of about 30. All the curves in Fig. 6.7 are now very close together and the effect of increased time of fouling is now minimal. Under these conditions the diffusional restriction is extremely severe and fouling effects are now of secondary importance.

Results of series fouling are shown in Fig. 6.9 in the form of a plot of η against ϕ for various values of the dimensionless time τ for the case of $b_o = 0$ and equal diffusivities of A and B. The dimensionless time for this case in the absence of parallel fouling is defined as $k_{f2}C_{Ao}t$. The effectiveness factor plots in Fig. 6.9 show only a slight decrease in particle effectiveness with fouling at ϕ values greater than unity, and there is very little difference in relative effectiveness between fresh and deactivated catalyst with increasing Thiele modulus. For a series mechanism, fouling is proportional to the concentration of product while the rate of reaction is proportional to the reactant concentration. Thus the fouling effect is greatest for high values of the Thiele modulus. This is because the reactant concentration drops to zero in the pellet interior under these conditions whereas the product concentration becomes very large. The activity plots within the catalyst pellet are shown in Fig. 6.10. At low values of the Thiele modulus the decrease in activity

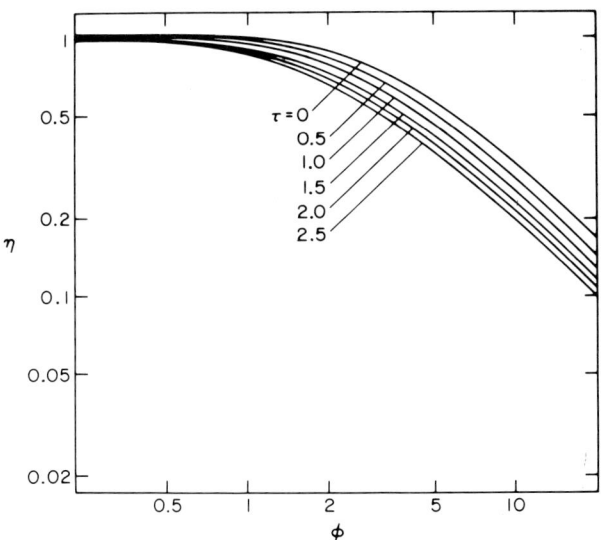

Fig. 6.9 Plot of η vs. ϕ for series fouling. (Kam et al., 1975)

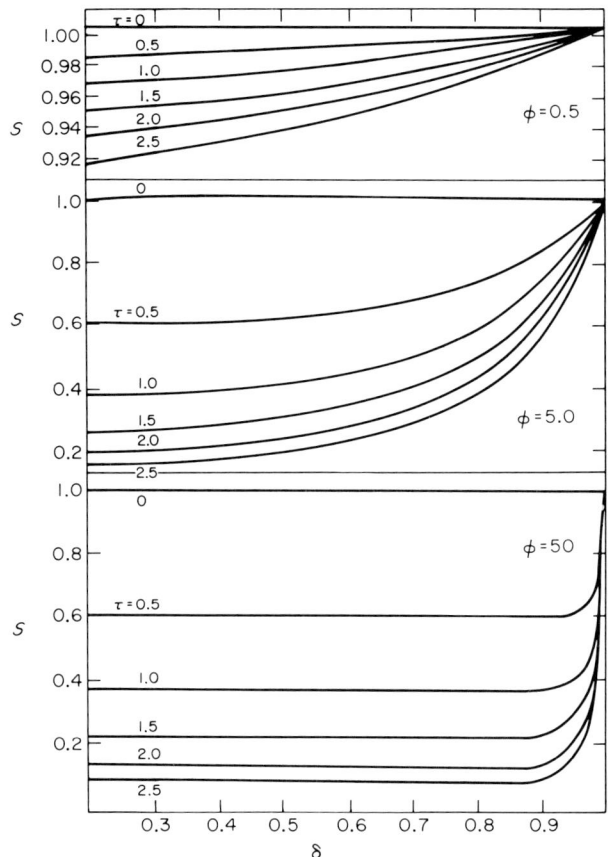

Fig. 6.10 Active sites distribution for series fouling. (Kam et al., 1975)

towards the pellet centre is quite small [Fig. 6.10(a)] but for larger values of ϕ the fouling is more pronounced [Fig. 6.10(b) for $\phi = 5.0$] and ultimately results in uniform fouling over a large portion of the interior volume of the catalyst while the activity at the surface of the pellet is unity [Fig. 6.10(c)]. A comparison of Figs 6.10(a) and 6.10(b) with similar plots for low ϕ values in parallel fouling [Figs 6.8(a) and 6.8(b)] shows that under these conditions fouling is much less pronounced for a series mechanism than for a parallel process. The large decrease in activity for series fouling at $\phi = 5.0$ is obscured by the decreased effectiveness due to pore diffusion limitations (Fig. 6.9), so the net effect is minimized. All the results quoted above have assumed a feed of 100% of reactant A with no product B present in the gas phase. In practice, some product would be expected to occur either from

6 Catalyst Deactivation by Fouling

recycle or towards the exit of the bed in a packed bed reactor. The effect of a finite concentration of product (b_o) on the fouling characteristics for a series mechanism is shown in Fig. 6.11 for a b_o value of 0.25. For this case there is almost uniform drop in catalyst activity irrespective of the Thiele modulus. This is because the activity at the surface is no longer unity as for series fouling with no product in the gas phase. The finite concentration of the product at the surface causes considerable fouling in the regions of the pellet close to the surface and causes a loss in activity even when the Thiele modulus is less than unity.

Where fouling occurs by a combination of series and parallel processes, the deactivation depends on the relative magnitude of the series and parallel fouling rate constants, k_{f2} and k_{f1} respectively, in equations (6.25) and (6.31). Results for parallel/series fouling have been obtained for a parallel fouling rate constant twice that for series. The concentration of product is assumed to be zero at the surface of the pellet (i.e. $b_o = 0$). Since $k_{f2}/k_{f1} = 0.5$ the effectiveness factor curves are similar in shape to those for parallel fouling, as might be expected, but the magnitude of the maxima at large values of the fouling time, τ, is much reduced. At values of the Thiele modulus greater than about 10, fouling has a greater effect than for parallel fouling alone (Fig. 6.7).

From these results it is clear that parallel fouling is more important than

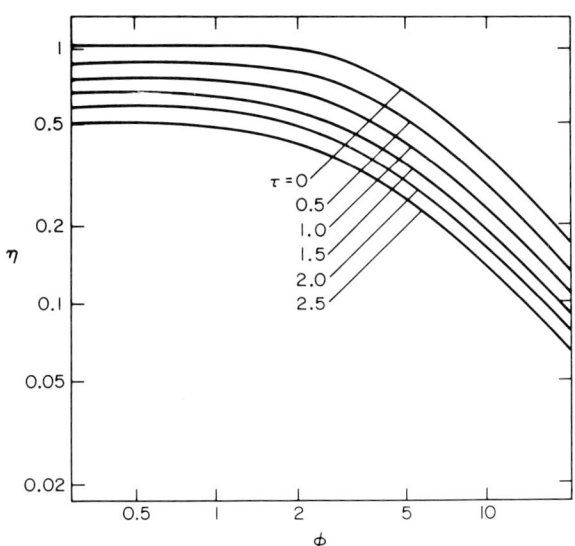

Fig. 6.11 Plot of η vs. ϕ for series fouling with initial product concentration ($b_0 = 0.25$) in bulk gas. (Kam et al., 1975)

series fouling for comparable values of k_{f1} and k_{f2}. This is because the parallel process depends on reactant concentration and this usually has a much greater driving force than series fouling. For low concentrations of product, series fouling only becomes important at high values of the Thiele modulus where the diffusional resistance is already of major importance. The only exception is when appreciable concentrations of product are present, e.g. if a large recycle ratio is employed or at the exit of a packed bed reactor.

The analysis has been extended to non-isothermal systems with the inclusion of film resistances. Two cases were considered, the first using the usual pseudo-steady state assumption (Kam et al., 1977a) and the second using a full transient analysis (Kam et al., 1977b). Again a Langmuir–Hinshelwood form of kinetics was assumed for the main reaction, while the fouling was assumed to occur by simple order kinetics as before. Because of the use of Langmuir–Hinshelwood kinetics the non-isothermal analysis must take account of the variation of adsorption with temperature, and therefore the heats of adsorption are included.

The concentration and temperature balance for the reactant A within the pellet can now be described by the following dimensionless equations:

$$\frac{d^2 a}{d\delta^2} + \frac{\alpha}{\delta}\frac{da}{d\delta} = \phi^2 f(a, \theta, S) \tag{6.39}$$

$$\frac{d^2 \theta}{d\delta^2} + \frac{\alpha}{\delta}\frac{d\theta}{d\delta} = -\beta\phi^2 f(a, \theta, S) \tag{6.40}$$

where $f(a, \theta, S)$ takes the form of the usual Langmuir–Hinshelwood expression for the main reaction A → B + C:

$$f(a, \theta, S) = \frac{aS \exp\left[(\gamma - h_{K_A})\left(1 - \frac{1}{\theta}\right)\right]}{1 + K_A^* \exp\left[-h_{K_A}\left(1 - \frac{1}{\theta}\right)\right]a + K_C^* \exp\left[-h_{K_C}\left(1 - \frac{1}{\theta}\right)\right]c} \tag{6.41}$$

where K_A^* and K_C^* are the dimensionless equilibrium absorption parameters and h_{K_A} and h_{K_C} the corresponding dimensionless heats of adsorption.

The deactivation relation combining both parallel and series fouling is now written as:

$$-\frac{dS}{d\tau} = aS \exp\left[\gamma_{f1}\left(1 - \frac{1}{\theta}\right)\right] + \frac{k_{f2}}{k_{f1}} bS \exp\left[\gamma_{f2}\left(1 - \frac{1}{\theta}\right)\right] \tag{6.42}$$

The initial and boundary conditions for the solution of these equations are:

$$\tau = 0; \quad S = 1 \text{ for any } \delta \qquad (6.43)$$

$$\delta = 0; \quad \frac{da}{d\delta} = \frac{d\theta}{d\delta} = 0 \text{ for any } \tau \qquad (6.44)$$

$$\delta = 1; \quad \frac{da_s}{d\delta} = \text{Sh}^*(1 - a_s); \quad \frac{d\theta_s}{d\delta} = \text{Nu}^*/(1 - \theta_s) \qquad (6.45)$$

Equations (6.39) and (6.40) may be simplified using the Prater relation which includes external resistances:

$$\theta = 1 + \beta \frac{\text{Sh}^*}{\text{Nu}^*}(1 - a_s) + \beta(a_s - a) \qquad (6.46)$$

This reduces the number of equations by one. Also, the effectiveness factor is now based on bulk conditions and is defined as:

$$\eta(\tau) = \frac{\alpha \int_0^1 \delta^{(\alpha - 1)} f(a, \theta, S) \, d\delta}{f(a_o, \theta_o)} \qquad (6.47)$$

In any non-isothermal simulation the values of the parameters selected are important. A general consensus puts the range of Thiele modulus as 0.5–45, while the thermicity factor β lies in the region 0.01–0.25 and the activation energy of the main and fouling reactions is in the range 2–65. The external heat and mass transfer parameters lie within $0.5 \leqslant \text{Nu}^* \leqslant 10$ and $20 \leqslant \text{Sh}^* \leqslant 50\,000$. For the adsorption parameters typical values are between 1 and 20 for K_A^* and K_C^*, while h_{K_A} and h_{K_C} may be assumed to be between 5 and 20. In the work of Kam et al. (1977a) values of $\gamma = 20$, $K_A^* = K_C^* = 10$, and $h_{K_A} = h_{K_C} = 5$ were employed, while β is taken as ± 0.02 depending on whether reaction was exothermic or endothermic.

In general, it was found that the effect of external mass transport on fouling was small for both isothermal and endothermic reactions. For the latter this is due to temperatures within the pellet being lower than those in the bulk stream, and reaction rates of both the main and fouling reactions are greatly reduced.

When exothermic reactions are considered with first order fouling kinetics, the value of the activation energy parameter for fouling, γ_f, was shown to be very important. It was observed that the relative drop in effectiveness factor due to fouling was very large for small values of the Thiele modulus for parallel fouling. This results from the high concentration of reactant throughout the pellet for low values of ϕ. At large values of ϕ the reaction occurs over a small region near the surface of the pellet, and for parallel fouling the decrease in activity is confined to this region. Hence, the relative drop in activity is very small in this asymptotic region.

It has been found that the magnitude of the activation energy of the fouling reaction is a very important parameter in non-isothermal fouling. However, the effect depends markedly on the diffusional resistance. At low values of the Thiele modulus the value of γ_f is unimportant. This is because temperature gradients in the catalyst are almost non-existent in this region. At large values of the Thiele modulus, γ_f becomes very significant since this region corresponds to control by the film resistance where the catalyst temperature would be very large and significantly higher than the bulk gas temperature. However, the fouling process leads to a large reduction in catalyst temperature. This is illustrated in Fig. 6.12 for a ϕ value of 30, which is clearly in the diffusion controlled region, and for γ_f values of 0, 5, and 10. The dimensionless surface temperature at $\tau = 0$ is 1.9331, while at $\tau = 0.4$ it drops to 1.750, 1.886, and 1.927 for $\gamma_f = 10$, 5, and 0 respectively.

The influence of γ_f can be more clearly seen by plotting the effectiveness factor as a function of time for a given ϕ (Fig. 6.13). These curves are for parallel fouling, and it can be seen that for $\phi = 1$ the curves for $\gamma_f = 5$ and $\gamma_f = 10$ almost coincide and are slightly below that for $\gamma_f = 0$, showing that γ_f has little effect at small values of ϕ. However, γ_f does become very significant at large values of ϕ and for $\gamma_f = 10$ where the effectiveness factor drops very sharply at a dimensionless fouling time of 0.1–0.2.

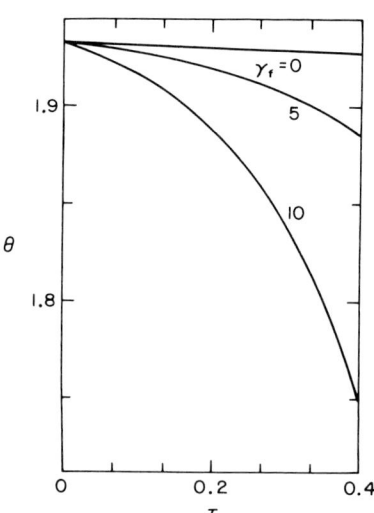

Fig. 6.12 Variation of particle temperature with process time. Effect of γ_f, parallel fouling; Sh* = 250, Nu* = 5, $\beta = 0.02$, $\gamma = 20$, $\phi = 30$. (Kam et al., 1977)

6 Catalyst Deactivation by Fouling

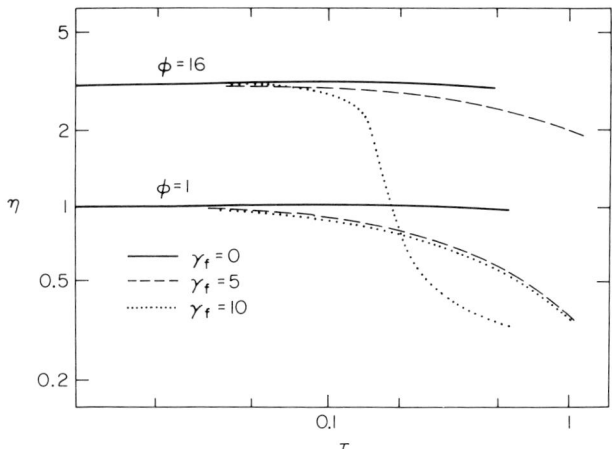

Fig. 6.13 Variation of effectiveness factor with process time for parallel fouling; Sh* = 250, Nu* = 5, $\beta = 0.02$, $\gamma = 20$. (Kam et al., 1977)

The effect of fouling mechanism on effectiveness factor is best illustrated by plotting the latter quantity against the dimensionless process time τ for different values of the Thiele modulus. Values of ϕ equal to 1 and 16 were chosen as representing the two extremes, and the results for series fouling, parallel fouling, and combined series/parallel fouling are given in Fig. 6.14. It is seen that for $\phi = 1$ parallel fouling causes a bigger decrease in catalyst

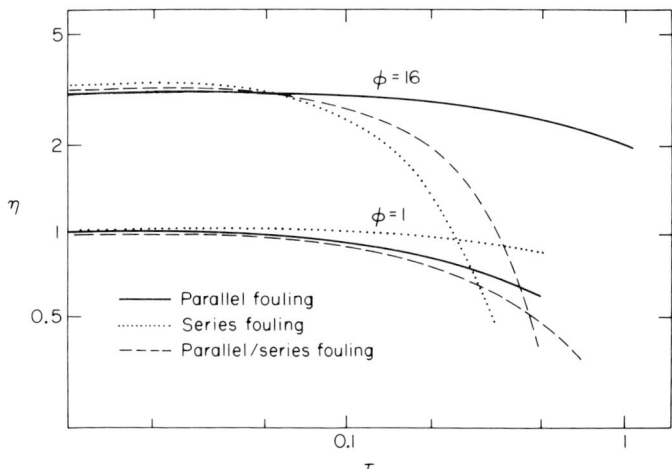

Fig. 6.14 Variation of effectiveness factor with process time. Effect of fouling mechanisms. Sh* = 250, Nu* = 5, $\beta = 0.02$, $\gamma = 20$, $\gamma_f = 5$. (Kam et al., 1977)

activity while for $\phi = 16$ series fouling causes more deactivation. Series/parallel fouling lies between these two extremes as expected. The curves in Fig. 6.14 are in contrast to those obtained previously by Masamune and Smith (1966) for isothermal fouling and by Sagara *et al.* (1967) neglecting film resistances. In both of these the curves of η against time intersected only for parallel fouling. The results in Fig. 6.14 show that intersections also occur for series fouling when external resistances are included.

In the transient analysis developed by Kam *et al.* (1977b) some interesting conclusions on the nature of fouling were found. Two transient models were used, one in which temperature gradients were allowed for in the pellet and the other a lumped isothermal pellet model in which all the temperature change was localized in the gas film surrounding the pellet. The latter was shown to give misleading results and therefore could not be employed under the conditions used. The extent of fouling compared with the main reaction was correlated by a characteristic parameter for fouling, σ, defined as:

$$\sigma = \frac{D_{eA}}{\varepsilon R^2 k_{f1} C_{Ao}} \tag{6.48}$$

It is a measure of the rate of intraparticle diffusion compared with the rate of fouling. Ranges of this parameter for practical reactions range from close to unity for very fast fouling to 10^3 or greater for slow fouling processes. When the characteristic parameter for fouling is very large the pseudo-steady state assumption is applicable, and thus use of this parameter enables a quick check to be made on the validity of this assumption for the particular system used.

In the above work the fouling reaction has been assumed to be of simple order even though the main reaction obeys Langmuir–Hinshelwood kinetics. It would seem more logical to assume that, if the main reaction has this form, which may involve adsorption of the reactant and product species, the fouling reaction should also be of Langmuir–Hinshelwood kinetic form. Studies using Langmuir–Hinshelwood kinetics for fouling have been made for isothermal systems by Chu (1968) and for non-isothermal systems by Kam and Hughes (1979).

An important factor when adsorption occurs with reaction is the magnitude of the adsorption constants. Chu suggested that an increase in the value of the appropriate adsorption constants would increase the rates of deactivation. This was confirmed by Kam and Hughes at values of ϕ equal to 0.3 and 3.0 for both parallel and series fouling, which is not altogether surprising since at these low ϕ values the catalyst temperature is almost equal to the bulk gas and near-isothermal conditions prevail. However, at larger values of ϕ a different effect was observed by Kam and Hughes. This is

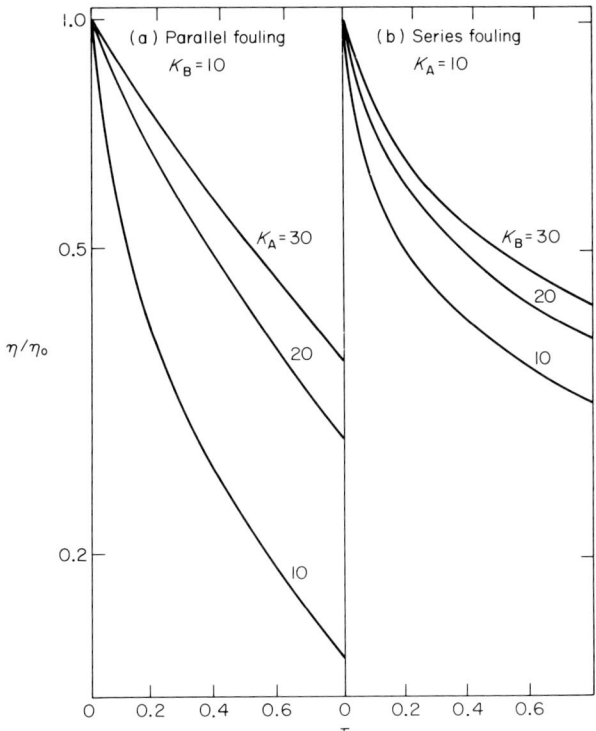

Fig. 6.15 Effect of K_A and K_B on relative effectiveness factor for (a) parallel and (b) series fouling; $\phi = 10.5$. (Kam and Hughes, 1981)

illustrated in Figs 6.15(a) and 6.15(b) for parallel and series fouling respectively for a value of ϕ equal to 10.5. At this ϕ value, the main reaction is under diffusion control and the reaction depends on the combined effects of adsorption equilibrium constant, concentrations, temperature, and activity of the catalyst. For parallel fouling K_B^* is fixed at 10 while K_A^* is varied in the range 10, 20, 30, while for series fouling K_A^* is fixed at 10 and K_B^* took values of 10, 20, and 30. For these simulations γ_f was 40 and γ was 20 while the activity is plotted as the ratio η/η_0, the relative effectiveness factor (η at τ divided by η at $\tau = 0$). Values of the surface temperature θ_s are given in Table 6.1. Since γ_f is greater than γ, fouling is favoured at higher temperatures.

The results in Table 6.1 show that for parallel fouling an increase in K_A^* is accompanied by a corresponding decrease in θ_s at this value of ϕ. This reduces the extent of fouling as shown in Fig. 6.15(a). For series fouling, the surface temperatures are also reduced but to a lesser extent than for parallel fouling. This would be expected to lead to more severe fouling in the series

Table 6.1 Particle surface temperatures.

Parallel fouling			Series fouling		
K_A^*	K_B^*	θ_s	K_A^*	K_B^*	θ_s
10	10	1.0708	10	10	1.0708
20	10	1.0156	10	20	1.0504
30	10	1.0065	10	30	1.0418

case, but the situation is complicated by the additional effect of K_B^* on the main reaction. A high value of K_B^* will hinder the adsorption of A, resulting in lower conversion of A into B which will, in turn, reduce the extent of fouling compared with the parallel case as shown in Fig. 6.16(b). It is interesting to note, however, that the greatest deactivation occurs when the surface temperature is greatest. Similar effects occur at values of ϕ greater than 10.5.

6.2 Catalyst deactivation due to deposition of impurities in the feed onto the catalyst

This type of fouling is chiefly found currently in hydrodesulphurization reactors used in process oil feedstocks in the petroleum industry. This frequently results in pore plugging caused by organometallic compounds present in the oil feedstock depositing as metal sulphides in the pore structure of the hydrotreating catalysts. The chief metal constituents giving this effect are iron, vanadium, and nickel. Similar reactions occur in the upgrading of coal liquids by hydrotreating although the relative proportion and type of metal sulphide deposited may be different. This pore plugging can result in a reduction of pore space and interstitial voidage in the packed bed resulting in an increase in reactor pressure drop as well as deactivation.

6.2.1 Modelling studies of pore plugging deactivation

The major work in this area has been carried out by Newson (1970, 1975) although Ozaki *et al.* (1975) have also estimated catalyst life based on kinetic studies and an analysis of the temperature gradient in a hydrodesulphurization reactor.

In his first paper Newson (1970) developed methods for estimating the pressure drop in the reactor due to bed plugging by metal sulphides deposition. In a later paper (Newson, 1975) he developed a theory for deactivation based on a pore plugging model.

The pore model of Wheeler (1951) was used, in which the pellet was

6 Catalyst Deactivation by Fouling

approximated as a composite of N pores each of length L_p where:

$$L_p = \frac{V}{S}\sqrt{2} \tag{6.49}$$

$$N(r) = \frac{S'\varepsilon(r)}{r^2\sqrt{2}} = \frac{K_1\varepsilon(r)}{r^2} \tag{6.50}$$

in which S' is the external surface area of the catalyst pellet and $\varepsilon(r)$ is the pellet porosity at a radius r.

The nature of the pore plugging process suggests that the number of pores effective for demetallation, N_E, is some fraction of the total given by:

$$N_E = \eta N \tag{6.51}$$

where η is the time averaged effectiveness factor based on metal profile analysis in spent catalyst.

The pore size distribution may be assumed to be Maxwellian:

$$L_M(r) = A_M \frac{r}{r_o} e^{-r/r_o} \tag{6.52}$$

where $L_M(r)$ is the total length of pores of radius r per gram of catalyst and A_M is the frequency of pore sizes, while r_o is the most probable pore radius and is linked to the average pore radius, $\bar{r}$, by:

$$\bar{r} = 3r_o = \frac{2V_g}{S_g} \tag{6.53}$$

The corresponding pore volume distribution is:

$$\varepsilon_M(r) = \int \pi r^2 L_M(r)\,dr \tag{6.54}$$

The deactivation rate in a single pore is then considered using the classical treatment of Wheeler (1951) giving the differential equation:

$$\pi r^2 D_e \frac{d^2C}{dx^2} = 2\pi r k C(x) \tag{6.55}$$

When the concentration gradient tends to zero at some distance $x_p \ll L_p$, equation (6.55) simplifies to:

$$\pi r^2 D_e \left(\frac{dC}{dx}\right)\Big|_{x=0} = 2\pi r k \int_0^{x_p} C(x)\,dx \tag{6.56}$$

where x_p is the reactant penetration into the pore. Now $(dC/dx|_{x=0})$ can be approximated by C_s/x and the reaction rate on the pore wall by $2\pi r k(C_s/2)$.

Equation (6.56) then becomes:

$$\pi r^2 D_e \frac{C_s}{x_p} = 2\pi r x_p k \frac{C_s}{2} \qquad (6.57)$$

Therefore

$$x_p = \sqrt{\left(\frac{D}{k} r\right)} = K\sqrt{r} \qquad (6.58)$$

Assuming a pseudo-steady state (a slow deactivation rate compared with oil residence time), the decrease in pore radius due to metal deposition may be determined. Thus, if the plugging material is spread over all the pores in a pore mouth type of plugging, the thickness of deposit, $y_{t,\Delta}$, for a finite increment of time Δ is given by:

$$y_{t,\Delta} = \frac{(R_p^*)_t \Delta}{2\pi \sum_m N_{E,m} r_{m,t} x_{m,t}} \qquad (6.59)$$

where R_p^* is the pore plugging rate with:

$$r_{m,t} = r_{m,0} - \sum_0^t y_t \qquad (6.60)$$

and

$$x_{m,t} = K\sqrt{r_{m,t}} \qquad (6.61)$$

for the mth pore at a time t.

Since the intraparticle metal deposition rate equals the flux of metal containing molecules in the pellet

$$R_D^* = F_t \alpha \sum_m \frac{N_{E,m}(r_{m,t})^2}{x_{m,t}} \qquad (6.62)$$

where R_D^* is the deposition rate. Also F_t/F_0 represents the ratio of the flux of metal containing molecules into the pellet at time t compared with that at zero time and is a measure of the decrease in reaction rate due to pore plugging.

The required relation between deactivation and pore plugging rate is obtained by assuming that deactivation is proportional to the rate of reaction and both are given by power law expressions of order 2:

$$Y \propto (R_D^*)^2 \propto (R_{HDS}^*)^2 \qquad (6.63)$$

where R_D^* is the deposition rate and R_{HDS}^* is the rate of hydrodesulphurization. Then

$$\frac{Y_0}{Y} = \left(\frac{R_{D,0}^*}{R_{D,t}^*}\right)^2 = \left(\frac{F_0}{F_t}\right)^2 \qquad (6.64)$$

represents the number of times the initial deactivation rate is increased. F_0 is known from fresh catalyst properties and process conditions; F_t can be calculated for any process time.

The procedure used was to calculate deactivation versus time using equation (6.64). The total pore plugging rate is then put into (6.59) to determine the reduction in pore radii over the whole pore size distribution using equation (6.52). The decrease in radii reduces the flux of metal containing models into the pores; equation (6.62). This flux is then compared with the initial value to calculate catalyst deactivation versus time.

Newson (1975) made predictions of catalyst life under conditions where metal deposition from feedstocks could occur. The predictions were compared with data from the operation of commercial plants; good order of magnitude agreement was obtained. This method should also be valuable in calculating restricted diffusion due to pore plugging in many other processes.

References

Beranek, L. (1972). *J. Catal.* **27**, 151.
Bischoff, K. B. (1976). Notes presented at NATO Sponsored Advanced Study Institute. Analysis of Fluid Solid Systems.
Chu, C. (1968). *Ind. Eng. Chem. (Fund.)* **1**, 509.
Eberly, P. E., Kimberlin, C. N., Miller, W. H. and Drushel, H. V. (1966). *Ind. Eng. Chem. (Proc. Des. Devel.)* **5**, 193.
Froment, G. F. (1976). Proc. 6th Int. Congr. on Catalysis (London), p. 10.
Froment, G. F. and Bischoff, K. B. (1961). *Chem. Eng. Sci.* **16**, 189.
Haldeman, R. C. and Botty, M. C. (1959). *J. Phys. Chem.* **63**, 489.
Kam, E. K. T., and Hughes, R. (1979). *A.I.Ch.E.J.* **25**, 359.
Kam, E. K. T., Ramachandran, P. A. and Hughes, R. (1975). *J. Catal.* **38**, 283.
Kam, E. K. T., Ramachandran, P. A. and Hughes, R. (1977a). *Chem. Eng. Sci.* **32**, 1307.
Kam, E. K. T., Ramachandran, P. A. and Hughes, R. (1977b). *Chem. Eng. Sci.* **32**, 1317.
Masamune, S. and Smith, J. M. (1966). *A.I.Ch.E.J.* **12**, 384.
Murakami, Y., Kobayashi, T., Hattori, T. and Masuda, M. (1968). *Ind. Eng. Chem. (Fund.)* **7**, 599.
Newson, E. J. (1970). Preprints A141–152, Div. Petrol. Chem. 160th National Meeting Amer. Chem. Soc. (Chicago, Sept. 13–18th).
Newson, E. J. (1975). *Ind. Eng. Chem. (Proc. Des. Devel.)* **14**, 27.
Noda, H., Tone, S. and Otaoke, T. (1974). *J. Chem. Eng. Japan* **7**, 110.
Ozaki, H., Satomi, Y. and Hisamitsu, T. (1975). 9th World Petrol. Congr. (Tokyo) **PD18**(4).

Ozawa, Y. and Bischoff, K. B. (1968). *Ind. Eng. Chem. (Proc. Des. Devel.)* **7**, 72.
Pachovsky, R. D., Best, D. A. and Wojciechowski, B. W. (1973). *Ind. Eng. Chem. (Proc. Des. Devel.)* **12**, 254.
de Pauw, R. P. and Froment, G. F. (1975). *Chem. Eng. Sci.* **30**, 789.
Ramser, J. H. and Hill, P. B. (1958). *Ind. Eng. Chem.* **50**, 117.
Richardson, J. T. (1972). *Ind. Eng. Chem. (Proc. Des. Devel.)* **11**, 8.
Sagara, M., Masamune, S. and Smith, J. M. (1967). *A.I.Ch.E.J.* **13**, 1226.
Villadsen, J. V. and Stewart, W. (1967). *Chem. Eng. Sci.* **29**, 1500.
Voorhies, A. (1945). *Ind. Eng. Chem.* **37**, 318.
Weekman, V. W. and Nace, D. M. (1970). *A.I.Ch.E.J.* **16**, 397.
Wheeler, A. (1951). *Adv. Catal.* **3**, 250.
Wheeler, A. (1955). In *Catalysis*, Vol. II (ed. P. H. Emmett).
Wojciechowski, B. W., John, T. M. and Pachovsky, R. A. (1974). *Adv. Chem. Ser.* **133**, 422.
Wojciechowski, B. W. (1974). *Catal. Rev. Sci. Eng.* **9**, 79.

7
Deactivation in Catalytic Reactors

7.1 Introduction

Several types of reactors may employ catalysts and therefore may be subject to catalyst deactivation. The common types used include fixed bed reactors, fluidized bed reactors, moving bed reactors, and slurry reactors. All of these may suffer from catalyst deactivation according to the type of reaction, the feed purity, etc. The first three types of reactor operate under steady state (or more generally pseudo-steady state conditions, if slow deactivation occurs) whereas slurry reactors used for processing gas–liquid systems with solid suspended catalyst particles, for example the hydrogenation of oils, operate in the batch or semi-batch mode.

The fluidized bed reactor and moving bed reactor represent systems where the solid catalyst is transported in the system. Therefore these types of reactor can easily have continuous regeneration facilities incorporated into the system by means of a reactor–regenerator sequence. Alternatively, if regeneration is not feasible for any given system, the used catalyst may be discarded and fresh catalyst added for these two types of reactor. A comparison of reactor types is made in the following chapter when optimization of catalyst deactivation problems is considered, and therefore discussion of these reactor types is deferred until then. The fixed bed reactor still represents one of the major areas of catalyst application, and when deactivation occurs in fixed beds these have to be taken off stream either for the catalyst to be discarded and a fresh catalyst charge inserted, or for regeneration of the existing spent catalyst to be attempted.

Fixed bed reactors, which consist of solid catalyst particles through and around which the fluid stream passes, represent a complicated physical and chemical system. The performance of a fixed bed reactor may be predicted by modelling but the agreement consequently obtained depends to a large extent on the degree of sophistication that may have to be introduced into the model.

The simplest assumption on flow behaviour which is frequently made is that flow occurs by a "plug" or "piston" like motion. This implies that a constant residence time exists for all fluid elements in the reactor, and this in turn represents the most efficient utilization of the reactor. In practice, plug flow is closely approximated in many industrial fixed bed reactors where the aspect ratio of length to tube diameter is high. Deviations from plug flow occur due to axial dispersion but for many purposes this can be ignored for all except shallow beds. Radial dispersion may be significant in certain instances but is normally of less importance than axial dispersion.

Temperature gradients in non-isothermal systems have a profound influence on reactor performance. Axial gradients will inevitably be present when reaction occurs due to the finite enthalpy of reaction. Radial gradients will be present if heat removal occurs at the reactor wall, but will not occur if the reactor is operated adiabatically. These temperature effects can usually be incorporated in the model albeit at the expense of time consuming operations.

Two basic approaches have been made to the problem of modelling fixed bed catalytic reactors. One considers the reactor as being composed of both catalyst particle and fluid phases, and because the heterogeneous nature of the reactor is fully accounted for, such a model is termed a heterogeneous model. The other does not recognize the heterogeneous nature of the reactor explicitly but assumes the system to be homogeneous with average properties assumed for the bed as a whole; such a model is termed a pseudo-homogeneous model. In general, pseudo-homogeneous models can often be employed; since the reactor contents are considered as a whole, only single mass and heat balances are necessary to define the system. Heterogeneous models are required, however, when the system is very sensitive to temperature, since under these conditions significant gradients of concentration and temperature may occur between the fluid and the solid surface of the catalyst particles. Therefore heat and mass transfer effects between fluid and solid phases are incorporated into the equations. The heterogeneous model represents a closer approach to reality but the computational effort is much greater because separate balances have now to be written for both solid and fluid phases. Both types of model with various degrees of sophistication (or with various simplifying assumptions) have been used in studies of catalyst deactivation. These are considered at appropriate points in the following discussion.

Although a division into the various deactivation processes is not so important when the reactor as a whole is considered, rather than the individual catalyst pellets, there is still some justification for continuing this classification. Thus, whilst reactor performance, and in particular the distribution of the deactivated zones in a fixed bed reactor, may be similar for impurity poisoning and parallel fouling due to coking, there are sufficient differences observed, especially in the few industrial plant studies reported, to

7 Deactivation in Catalytic Reactors

justify retention of this classification. Accordingly, poisoning studies of fixed bed reactors will be considered next, followed in turn by coking studies, and finally by the few reported works on sintering in fixed beds.

7.2 Poisoning in fixed bed reactors

7.2.1 Isothermal analysis

One of the earlier systematic studies of the likely effects due to poisoning in fixed beds was made by Anderson and Whitehouse (1961). They considered the effect of various deactivation expressions of the type listed in Chapter 3 and fitted these to various types of concentration distribution equations. From these an average activity for the fixed bed was obtained. Essentially, the work involved the examination of the effect of these empirical deactivation expressions on reactor performance. Deactivation expressions of the form:

$$S = S_o - b_1 C_p \tag{7.1}$$

$$S = S_o \exp(-b_2 C_p) \tag{7.2}$$

$$\frac{S}{S_o} = \frac{1}{(1 + b_3 C_p)} \tag{7.3}$$

$$\frac{S}{S_o} = (1 - b_4 C_p)^{\frac{1}{2}} \tag{7.4}$$

were all examined and plots of residual average activity in the bed against poison concentration were obtained. These were linear for a wide range of poison concentration, thus justifying the Voorhies type of relation frequently obtained.

A more detailed approach to the problem of deactivation by impurity poisoning in packed beds was made by Wheeler and Robell (1969) who used the theory of adsorption in fixed beds to account for the poison distribution. The starting point of the analysis was the single pellet curves of activity remaining vs. fraction of surface covered, originally produced by Wheeler (1955). These curves can be represented algebraically by the expression:

$$\frac{k}{k_o} = \frac{S}{S_o} = \left(\frac{1}{1 + \phi_o C_p/C_{p\infty}} - \frac{C_p/C_{p\infty}}{1 + \phi}\right) \tag{7.5}$$

where ϕ_o is the Thiele modulus for the unpoisoned catalyst, C_p is the poison concentration on the surface, and $C_{p\infty}$ the saturation poison concentration ($S/S_o = 0$ when $C_p = C_{p\infty}$).

The distribution of the poison in the bed is obtained from the adsorption theory of Bohart and Adams (1920) applicable to fixed bed adsorption, which gives the adsorption rate as:

$$\frac{dC_p}{dt} = r_{ads} = k_{ads} C_{p,g}(1 - C_p/C_{p\infty}) \tag{7.6}$$

and if an isothermal bed and plug flow are assumed:

$$\frac{dC_{p,g}}{dx} = r_{ads} \tag{7.7}$$

In the above, k_{ads} is the adsorption rate constant and $C_{p,g}$ is the gas phase poison concentration.

The solution for a bed initially free of poison is:

$$\frac{C_p}{C_{p\infty}} = \frac{1 - \exp(-N_t\theta/\theta_\infty)}{1 + \exp(-N_t\theta/\theta_\infty)[\exp(N_t x/L) - 1]} \tag{7.8}$$

where N_t is equal to $k_{ads}L/v$ and represents the number of adsorption transfer units in the reactor, and θ_∞ is the ratio of the total capacity of the catalyst for adsorption of poison to the rate of poison admission to the reactor:

$$\theta_\infty = \rho_B C_{p\infty} L / MvC_{p,g}^o$$

where ρ_B is the catalyst bulk density, M the molecular weight of poison, and $C_{p,g}^o$ the poison inlet concentration.

Equations (7.5) and (7.8) can be combined and the plug flow reactor equation for reactant conversion can then be evaluated analytically. The general solution obtained for the exit conversion of reactant as a function of time on stream is:

$$\ln\left(\frac{C_A}{C_{Ao}}\right) = -\frac{k_o/k_{ads}}{1 + k_o} \left\{ \ln\left[1 + \exp\left(-N_t\frac{\theta}{\theta_\infty}\right)(\exp N_t - 1)\right] \right.$$

$$\left. + \phi_o \ln\left[1 + \frac{\exp\left(-N_t\dfrac{\theta}{\theta_\infty}\right)(\exp N_t - 1)}{1 + \phi_o\left[1 - \exp\left(-N_t\dfrac{\theta}{\theta_\infty}\right)\right]}\right] \right\} \tag{7.9}$$

For completely homogeneous poisoning, $\phi_o \to 0$ and equation (7.9) becomes:

$$\ln\left(\frac{C_A}{C_{Ao}}\right) = -\left(\frac{k_o}{k_{ads}}\right) \ln\left\{1 - \exp\left(-N_t\frac{\theta}{\theta_\infty}\right) + \exp\left[N_t\left(1 - \frac{\theta}{\theta_\infty}\right)\right]\right\}$$

$$\tag{7.10}$$

7 Deactivation in Catalytic Reactors

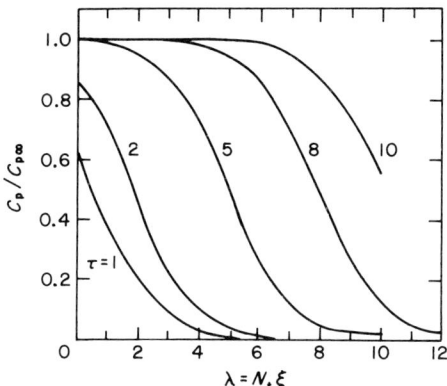

Fig. 7.1 Bohart–Adams generalized poisoning wave profiles. Plots of $C_p/C_{p\infty}$ against $\lambda = N_t \xi$ for various reduced times, $\tau = N_t \theta/\theta_\infty$. (Wheeler and Robell, 1969)

Typical poison profiles obtained by Wheeler and Robell (1969) are shown in Fig. 7.1 in which dimensionless poison concentration $C_p/C_{p\infty}$ is plotted against $N_t(l/L)$ for various reduced times τ given by $\tau = N_t(\theta/\theta_\infty)$. Effectively, this is a plot of poison concentration profiles for different times on stream and demonstrates the development of a "poison wave" in the bed which passes through the reactor at constant velocity and with a fixed shape. This type of behaviour is well documented in many industrial processes where catalyst deactivation by impurity poisoning is occurring. Wheeler and Robell themselves obtained experimental confirmation for their analysis in a study of the poisoning by H_2S of a 1%Pt–1%Pd on alumina catalyst used for the oxidation of CO. The results were plotted in the form of reactant converted against the time on stream and are shown in Fig. 7.2. The curve represents the calculated results while the experimental results are shown as circles. The parameters of the model were estimated by assuming homogeneous poisoning with a low value of the Thiele modulus, corresponding to curve A of Wheeler's classification (Fig. 3.5 in Chapter 3). It should be noted that four parameters are needed to fit this model; these are θ_∞, N_t, k_o/k_{ads}, and ϕ_o. Although the model requires estimation of these parameters the fit is still seen to be very good and other models often require an estimate of the same number of parameters.

The effect of diffusivity on poisoning in fixed bed reactors is of practical importance, if diffusivity is of significance in influencing poisoning, since it offers a means whereby the catalyst may be made more resistant to poisoning by adjusting the average pore size of the catalyst. Surprisingly little direct work has appeared on this topic but Olson (1968) has studied this effect for

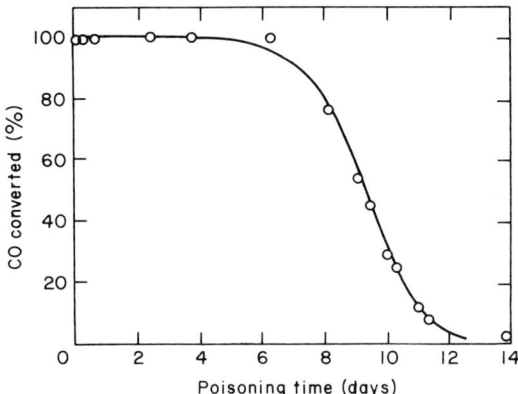

Fig. 7.2 Comparison between calculated and experimental catalytic activity history for CO oxidation at 25°C. Curve is calculated from equation (7.9) and points are experimental. Catalyst 1% Pt–1% Pd supported on alumina. Average H_2S concentration 13 ppm; CO concentration 83 ppm. (Wheeler and Robell, 1969)

the special case of Knudsen diffusion. This type of diffusion is directly proportional to the pore diameter and therefore can be changed by altering the pore size of the catalyst. In Olson's work it was assumed that the poison occurred as an impurity in the feed and accumulated on the catalyst at the entrance to the reactor to a greater extent than at the exit by a "shell progressive" mechanism. The bed thus develops variable pore mouth poisoning throughout its length.

For the bed a balance based on the adsorption equation derived by Vermeulen (1958) for ion exchange systems was used. By numerically integrating this by the method of characteristics the following dimensionless equation was obtained:

$$\frac{\partial C'_p}{\partial \xi} + \frac{\partial Q}{\partial \tau} = 0 \quad (7.11)$$

where C'_p is the dimensionless gas phase poison concentration, Q the volumetric fraction of the solid phase saturated with poison, and ξ and τ are the dimensionless reactor length and time on stream respectively. The shell progression mechanism requires that the poisoned portion of the catalyst be completely saturated. Assuming a pseudo-steady state for the movement of the poisoned zone in a spherical isothermal pellet, the mass flux of poison through the gas film is equated to that through the spherical shell of radius R and with the rate of chemisorption of poison at this interior radius R. The rate of decrease of R is then:

7 Deactivation in Catalytic Reactors

$$-\frac{q_s}{bC_{po}} R^2 \frac{dR}{dt} = \frac{k_c}{r_p} (C'_{pB} - C'_{pS})$$

$$= \frac{D_K}{r_p^2} (C'_{pS} - C'_{pI}) \frac{R}{1-R} = \frac{k_p}{r_p} R^2 C'_{pI} \quad (7.12)$$

where q_s is the saturation concentration of poison in the solid phase, r_p is the pellet radius, and the subscripts B, S, and I refer to bulk, surface, and interface concentrations of poison respectively. The unknown concentrations C'_{pS} and C'_{pI} may be eliminated from equation (7.12) and the dimensionless solid concentration Q and dimensionless shell radius R are related by:

$$Q = 1 - R^3$$
$$-dQ = 3R^2 \, dR \quad (7.13)$$

Introducing these and the dimensionless time into equation (7.12) we obtain:

$$\frac{dQ}{d\tau} = \frac{\left(\frac{1-\varepsilon}{\varepsilon}\right) N_s C'_{pB}}{\left[\frac{1}{Sh^*} + \frac{1-(1-Q)^{1/3}}{(1-Q)^{1/3}} + \frac{1}{D_a(1-Q)^{2/3}}\right]} \quad (7.14)$$

where $N_s = (3D_K L/r_p^2 u)$ is the number of solid diffusion transfer units in the bed, $Sh^* = (k_c r_p/D_K)$ is the modified Sherwood number for mass transfer, and $D_a(k_p r_p/D_K)$ is the Damköhler number for the poisoning reaction.

Boundary conditions for equations (7.11) and (7.14) are:

$$Q(0, \xi) = 0$$
$$C'_p(\tau, 0) = 1 \quad (7.15)$$

If a first order irreversible reaction is assumed in the catalyst, the effectiveness factor can be written as:

$$\eta = \frac{3}{\phi^2 \left[\frac{1}{Sh_r^*} + \frac{1-R}{R} + \frac{1}{R(R\phi \coth(R\phi) - 1)}\right]} \quad (7.16)$$

where $\phi = (k_r r_p^2/D_K)^{\frac{1}{2}}$ is the Thiele modulus for the main reaction and Sh_r^* is the modified Sherwood number for the main (catalytic) reaction. The effect of poisoning on the overall activity of the bed with linearity of the assumed kinetics permits simple averaging:

$$\bar{\eta}(\tau) = \int_0^1 \eta(\xi, \tau) \, d\xi \quad (7.17)$$

where $\bar{\eta}(\tau)$ is the average effectiveness. The fractional activity ratio, representing the activity that remains after poisoning to the initial activity, is written as:

$$S(\tau) = \frac{\bar{\eta}(\tau)}{\eta(0)} \qquad (7.18)$$

and this is the quantity of prime interest.

The results obtained using this analysis (which it should be remembered assumes isothermal conditions) are shown in Fig. 7.3 where the activity ratio is plotted against time on stream for three values of the Thiele modulus and two values of N_s. At a time equal to unity the reactor has been supplied with sufficient poison to completely poison the whole bed. For a Thiele modulus equal to 100 it can be observed that for times less than 0.4 the activity ratio for $N_S = 1.0$ is greater than that for $N_S = 0.1$. This unexpected result can be explained as follows. When ϕ is large the poison penetration into the catalyst is small, and when N_s is large the axial concentration profile of poison in the bed has a narrow band width. This means that a substantial portion of the bed is totally unpoisoned under these conditions. When N_S is small, however, the bed is poisoned more uniformly, resulting in a rapid reduction of the main reaction rate for systems sensitive to pore mouth poisoning. Thus, the

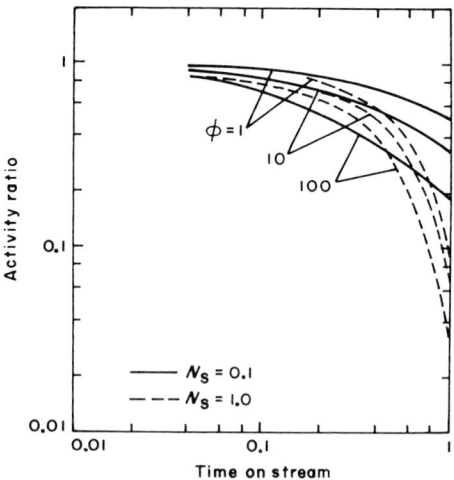

Fig. 7.3 Activity of bed. The fraction of the initial activity of the bed is plotted as a function of the time on stream. At time equal to unity, the reactor has been supplied with enough poison to saturate the bed completely. The parameter N_S describes the transport rate of poison into the pellet while the Thiele parameter refers to the major chemical reaction. $Sh^* = Da = 20$ (Olson, 1968)

two important factors are N_s and ϕ. Low values of N_S imply that the adsorption of poison occurs slowly, and this generally tends to affect reactor performance only slightly. However, if the Thiele modulus is large, a uniformly poisoned bed is less effective than one poisoned with a steep poison profile, i.e. with a large N_s value. Thus, for low values of the Thiele modulus, the useful lifetime of the reactor is improved by decreasing the Knudsen diffusivity, while at high values the converse is true.

Further results obtained by Olson showed that the radius of the unpoisoned core varied with the value of N_S for a time $\tau = 0.4$. Results are shown in Fig. 7.4 and confirm that for $N_S = 1.0$ the latter part of the bed is completely free of poison while the front part is substantially poisoned. However, for $N_S = 0.1$, a thin shell of poison is distributed throughout the whole bed length.

This type of analysis based on adsorption theory has been extended by other workers. Koch et al. (1980) used a similar analysis to that of Wheeler and Robell in a study of the liquid phase hydrogenation of a cyclic amine and found that their experimental results could be well represented by such an analysis. Haynes (1970) has also extended the shell progressive model using separate Thiele moduli for both reactant and poison. In all these analyses certain key parameters have to be estimated. In general these are the rate constant for the main reaction, the adsorption rate constant, the diffusivities of the poison, and/or the main reactant in the catalyst pore structure and the saturation capacity of the catalyst for the poison.

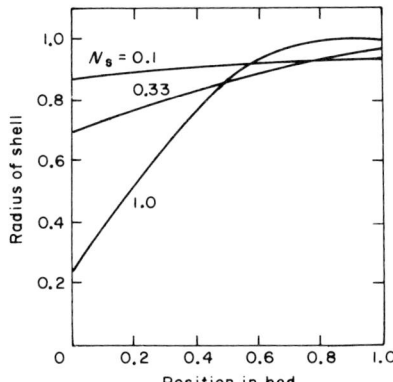

Fig. 7.4 Interfacial position of poisoned shell. The radial position of the poisoned shell (Fig. 7.3) is shown as a function of axial position at dimensionless time $\tau = 0.4$. The catalyst at the entrance to the bed is not completely poisoned for any value of the parameter N_S. Therefore the "constant band width" assumption is not valid. (Olson, 1968)

All the above have assumed the catalyst to be monofunctional. In fact, many industrial catalysts are bifunctional in nature, as in the platinum on alumina catalysts used in reforming naphtha. Lee and Butt (1973) have analysed the effect of poisons on the selectivity of bifunctional catalysts. While the analysis is complex and is not detailed here, the conclusions are interesting. It was found that some degree of diffusional limitation improved both catalyst efficiency and catalyst life. Thus, it would appear that even for this complex case a modest increase in diffusional resistance is beneficial, a conclusion generally in line with those reported for the simpler monofunctional catalysts.

7.2.2 Non-isothermal analysis

Although the progress in isothermal analysis of poisoning has led to a better understanding of the process of catalyst deactivation by impurity poisoning, the majority of reactions of industrial importance are not isothermal and steep temperature gradients can occur in the reactor. Therefore any complete analysis must account for the non-isothermal behaviour of the system.

If the reactor is used for processing an exothermic reaction and operates adiabatically, a steep temperature profile is frequently obtained if, as is usually the case, the catalyst is very active chemically. Because of this the reaction may be controlled by mass transfer across the gas film and the so-called "ignition" type temperature profile may be obtained. Under these conditions the reaction is confined to a narrow zone in the bed, usually only a few catalyst pellets in length. If poisoning occurs in this situation it too will be limited to this reaction zone, and as the zone becomes poisoned the reaction zone will pass down the bed. The temperature profiles indicate the position of the reaction zone and the displacement of this due to poisoning.

In practice, the usual quantities monitored in industrial reactors are the exit conversion and the temperature profile (sampling of the reactant concentration along the reactor is not always practical). The temperature profile therefore serves to estimate the residual activity of the reactor. Typical temperature profiles and their variation with time are shown in Figs 7.5(a) and 7.5(b), and these illustrate the two distinct ways in which reactor performance may deteriorate with time. In Fig. 7.5(a) the temperature profile "marches" down the bed, retaining essentially the same shape with time elapsed. As the front portion of the bed is poisoned the reaction zone moves into the next unpoisoned region, and this continues until the exit of the bed is reached. This behaviour is similar to the predicted analysis of Wheeler and Robell (1969) which was based on an isothermal adsorption of poison, and indeed such profiles are typical of the constant pattern profiles often exhibited in gas adsorption columns (Vermeulen, 1963). Many industrial systems

7 Deactivation in Catalytic Reactors

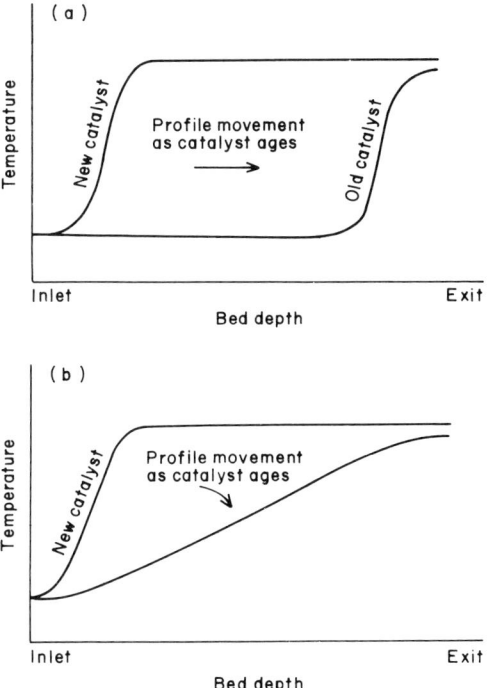

Fig. 7.5 Movement of temperature profile in a catalyst bed as the catalyst ages.

exhibit this type of poisoning behaviour, including the low temperature copper shift catalyst used in ammonia production units and the catalytic rich gas catalyst for methane production from low temperature steam reforming of petroleum fractions, both of which are well documented in the literature.

The behaviour depicted in Fig. 7.5(b) is somewhat different, in that the temperature profile "reclines" as the catalyst deactivates throughout the bed. Sometimes this deactivation may occur through poor stabilization of the crystallites in a supported metal catalyst or because of poisons that are not permanently adsorbed by the catalyst.

The ultimate effect from both types of profile, as shown in Fig. 7.5 (a) and (b), is that the product gas will not have achieved equilibrium before it leaves the reactor, and therefore "slip" of material will occur from the end of the bed. Where the reaction rate is limited by mass transfer of any sort, the surface area of the pellets exposed to the gas becomes important. Smaller pellets will tend to give faster rates, while with larger pellets the centre of the pellet may be masked and access to the reactants is limited. Thus, in general, small particles would be expected to give a higher rate and a sharper temperature profile but will also give a greater pressure drop in the bed; a compromise has therefore to be reached.

In contrast to the wealth of plant data, there have been very few systematic laboratory investigations reported of non-isothermal poisoning in fixed bed reactors. However, one very detailed and elegant study has been made by Weng et al. (1976) of the thiophene poisoned hydrogenation of benzene over a nickel–kieselguhr catalyst (12–20 mesh). This is essentially an example of irreversible poisoning, and in a comprehensive series of experiments rate data were obtained for the reaction under poison-free and poisoned conditions. The main reaction was found to be well fitted by a Langmuir–Hinshelwood rate expression which included a heat of adsorption term, while for the poisoning reaction a first order expression in both the gas phase poison concentration and the active sites concentration was found to represent the results adequately. Reactor temperature profiles were obtained as well as exit conversions. The temperature profiles were of the "marching" type depicted in Fig. 7.5(a), but since the reactor was not operating adiabatically these were in the form of temperature peaks or "hot spots" which travelled through the bed. The temperature peaks were of roughly constant height and in all cases 100% conversion of the limiting reactant (benzene) was achieved. These experimental profiles were simulated using an axial dispersion model, and reasonable agreement in the magnitude of the temperature peaks was found, but the velocity of propagation of the reaction zone was found to be about one-half of the experimental value. Possible causes for this discrepancy were considered, including (a) errors in the saturation poison capacity measurements which were part of the experimental programme and (b) the kinetic model for the poisoning reaction being really not a linear function of poison and active sites concentration. The latter was considered to be the probable source of the discrepancy and the poisoning process was modified to a dual site model. This was found to give much better agreement with experiment.

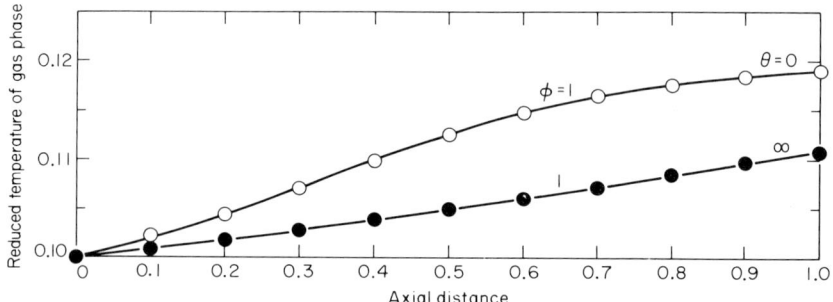

Fig. 7.6 Simulation of poisoning in fixed bed reactor. $J_H = 10$, $J_M = 200$, $\beta = 0.2$, $Da = 1$, $\gamma = 10$, $K_p = 1$.

7 Deactivation in Catalytic Reactors

As a consequence of their results the authors emphasize that poisoning kinetics may be more complex than hitherto supposed, and care should be taken in adopting simple deactivation rate expressions and also in assuming separability of main and deactivation reactions.

The case of non-isothermal poisoning for a reversible reaction was analysed by Valdman (1976) in a modelling study. For this study, a heterogeneous model of the catalyst bed was developed, assuming plug flow but with all the thermal resistance of the pellet lumped into the gas film so that the catalyst pellet was assumed to be isothermal at a temperature level above the surrounding gas (the main reaction was assumed to be exothermic). Carberry (1966) and McGreavy and Cresswell (1969) have shown that this assumption is valid for many catalytic systems. The main reaction was assumed to be of the form:

$$R^* = \frac{kC_A}{1 + K_P C'_P} \qquad (7.19)$$

where the term $(1 + K_P C'_P)$ takes into account the poisoning of the catalyst. This means effectively that the deactivation expression is of the form of equation (7.3) above. The reaction rate is then first order in the reactant and an analytical expression for the effectiveness factor of the isothermal pellet may then be used. The poison adsorption is assumed to be given by the simple Langmuir expression (5.26) used in Chapter 5.

For an adiabatic reactor operating with an exothermic main reaction, concentration and temperature profiles were calculated for the unpoisoned reaction and for the poisoned reaction at a time sufficient for the poison adsorption equilibrium to be achieved (t_∞). Results are shown for the reduced gas phase temperatures (T/T_0) against axial distance in Fig. 7.6 for a Thiele modulus of 1. Although a high value of β equal to 0.2 was chosen for this simulation, the profiles, even for the unpoisoned case, were shallow. On being poisoned the profiles "reclined" even further in the manner of those shown in Fig. 7.5(b). The reactant concentration rose throughout the bed following the onset of poisoning, as is to be expected.

7.3 Coking of catalytic reactors

There is an analogy between single catalyst pellets and catalytic reactors in the way the coke deposit is generally considered to be distributed. Thus, we have already seen in Chapter 6 that for single catalyst pellets the coke distribution depends on the mechanism of coking. For a simple reaction of the form A → B the coke may originate from side reactions either in parallel or consecutive to the main reaction:

A → Coke   (Parallel process)

B → Coke   (Consecutive or series process)

It was noted, for single pellets subject to internal diffusional restrictions, that for parallel fouling coke tends to deposit in the outer regions of the catalyst pellet where the reactant concentration is highest. Similarly, for series fouling coke tends to deposit preferentially in the centre of the pellet if the diffusional resistance is not too great.

If the concentration dependence is dominant, similar arguments would be expected to apply to fixed bed reactors. Thus, for coking by a parallel mechanism, deposition of coke would be expected to be greatest at the entrance region of the bed where the reactant concentration is greatest. For coking by a series mechanism, coke deposition should only become significant towards the end of the bed where concentrations of product build up. One difference between single pellets and fixed bed reactors is that, for the latter, product concentrations can become appreciable in the gas phase and therefore contribute to the overall driving force for series fouling, whereas for single pellets product distributions inside the pellet can become modified due to diffusional restrictions.

For simple kinetics and isothermal systems the above description of coke distribution seems to be valid, as demonstrated by the work cited below. However, in non-isothermal systems with complex kinetics, these effects may not be universally true.

7.3.1 Isothermal coking in fixed bed reactors

Most studies have concentrated on this type of system, partially because of the simplicity of the balances, although it should be noted that quite complex kinetics have been applied to this type of analysis.

The first detailed analysis of isothermal coking was given by Froment and Bischoff (1961, 1962). A simple plug flow homogeneous model was adopted with constant density of the gas phase, and isothermality was also assumed. For the main reaction the pseudo-steady state balance for A may be written in terms of the mole fraction, x, as:

$$\frac{dx}{dz} = -\frac{A_c \rho_B d_p}{F} r_A \qquad (7.20)$$

in which the dimensionless variables τ and z are defined as:

$$\tau = \frac{F}{\varepsilon \rho_A A_c d_p} \theta; \qquad z = L/d_p, \ Z = \xi/d_p$$

where F is the feed rate, ε the void fraction, d_p the particle diameter, A_c the reactor cross section, ρ_A and ρ_B the density of reactant A and bulk density of catalyst respectively, L the reactor length, θ the time, and r_A the rate of disappearance of A. For the rate of deactivation, which is assumed to be measured by the rate of accumulation of coke, we have:

$$\frac{dC_c}{d\tau} = \frac{\varepsilon \rho_A A_c d_p}{F} r_c \tag{7.21}$$

where C_c is the carbon content expressed as a weight fraction of the catalyst weight.

For parallel coking Froment and Bischoff proposed the following rate equations:

$$r_A = k'_A Px + k'_{Ad} Px \tag{7.22}$$

$$r_c = k'_{Ad} P(1-x) \tag{7.23}$$

where P is the total pressure.

For series coking the equations are:

$$r_A = k'_A Px \tag{7.24}$$

$$r_c = k'_{Ad} P(1-x) \tag{7.25}$$

The proportionality of both main and side reactions to the activity of the catalyst at any time was then fitted by a multiplying factor to the rate constant, called an activity function ψ. Two forms of activity function were used, an exponential form and a hyperbolic form, defined as follows:

$$\psi = \exp(-\alpha C_c) \tag{7.26}$$

$$\psi = (1 + KC_c)^{-1} \tag{7.27}$$

These activity functions will, in general, affect the rate of both the main and deactivation reactions. If there is no effect on the deactivation reaction (i.e. the coke has neither a positive nor a negative self catalytic effect), ψ for the deactivation reaction is equal to unity.

Carbon profiles were obtained as analytical solutions to equations (7.24)–(7.27) with the conservation equations and approximate boundary conditions. Carbon profiles for parallel and series fouling are shown in Fig. 7.7 (a) and (b) respectively for an exponential activity function. As can be seen, deposition of coke is greatest at the entrance of the reactor for parallel fouling where the reactant (coke precursor) has the higher concentration. Conversely for series fouling, deposition is greatest near the reactor exit where the fouling precursor (in this case the product) has the highest concentration. Furthermore, parallel fouling gives a descending coke profile,

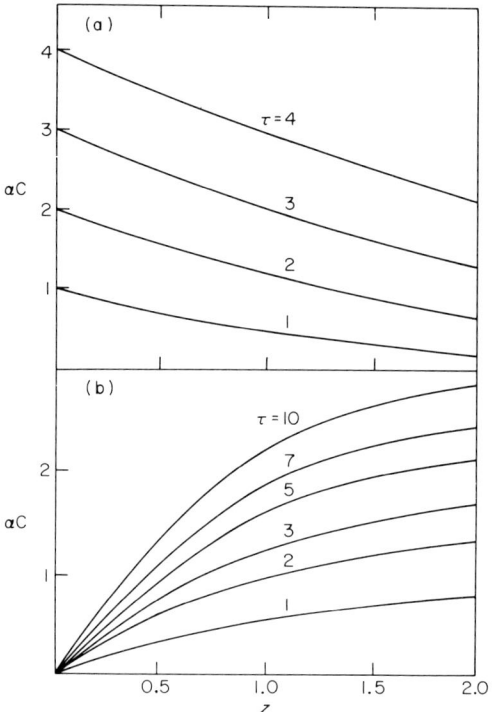

Fig. 7.7 Carbon profiles in a fixed bed reactor for (a) parallel fouling and (b) series fouling. (Froment and Bischoff, 1961)

while series fouling gives an ascending profile. Because of the activity relation (7.26), activities (more specifically, rate coefficients) will increase along the reactor with parallel fouling but will decrease with axial distance for series fouling. Similar results were also found with a hyperbolic activity function, and in this case the carbon forming reaction was also considered to be affected by catalyst fouling.

Experimental confirmation of coke profiles corresponding to parallel coking has been obtained by Van Zoonen (1965). The reaction considered was the hydroisomerization of olefins to paraffins over a silica–alumina–nickel sulphide catalyst in the temperature range 300–400°C with pressures of 40 bar. The reaction proceeds by a complex mechanism over this bifunctional catalyst; the silica–alumina causes the formation of isoparaffins and diolefins, and the latter are hydrogenated over the NiS back to the mono-olefin. Coke is formed from the olefin via the diolefin and its rate of formation is therefore proportional to the olefin concentration. The main reaction was taken as first order in reactant and inversely proportional to the

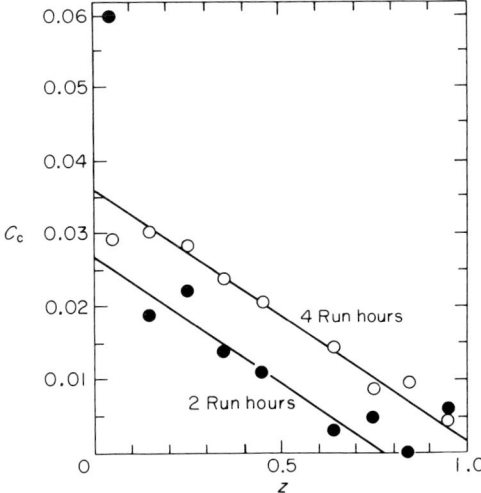

Fig. 7.8 Relation between coke content of catalyst, C_c (kg coke/kg catalyst), and location, z, in catalyst bed. 1-Hexene; atm. pressure; $S_v = 0.11$ kg kg^{-1} h^{-1}; H$_2$/olefin mole ratio = 4.7. (Van Zoonen, 1965)

coke concentration. This results in a set of equations similar to those used by Froment and Bischoff when isothermal conditions and plug flow are assumed. Results obtained for the hydroisomerization of 1-hexene are shown in Fig. 7.8 in which the coke profiles are illustrated for run times of 2 and 4 hours. Although there is some scatter, the decreasing coke profile with distance as a result of parallel fouling is clearly demonstrated. The continuous lines for each time are computed results using the model developed which was similar to that of Froment and Bischoff.

One of the difficulties in the Froment and Bischoff (1961, 1962) approach is of determining the correct form of the activity function. As noted above, Froment and Bischoff originally used both exponential and hyperbolic activity functions while Van Zoonen appears to have used a hyperbolic function alone to fit his results. The experimental work of Ozawa and Bischoff (1968a,b) on the coking of the ethylene hydrogenation reaction over a silica–alumina catalyst has been mentioned earlier in connection with the differences in surface area, pore size, and diffusivity consequent on coking a catalyst. The experimental data were also fitted empirically to a log-log plot of coke versus process time. In their second paper, which analysed the results of the experimental measurements, Ozawa and Bischoff (1968b) concluded that a linear deactivation function best described their results:

$$\psi = 1 - aC_c \tag{7.28}$$

A similar function was used in the modelling study of Masamune and Smith (1966). Additionally, Ozawa and Bischoff (1968b) showed that the constant a in equation (7.28) could be interpreted via a Langmuir–Hinshelwood mechanism, and the values of the rate constant for this deactivation function could be plotted on an Arrhenius diagram to give an activation energy close to that for the main reaction. Thus, under the conditions of their experiments, both the main and coking reactions would be equally affected by changes in operating temperature.

Ozawa and Bischoff also attempted to treat other deactivation data for coking, reported in the literature, in a similar manner. In particular, the results of Eberly *et al.* (1966) on the cracking of n-hexadecane over a silica–alumina catalyst were analysed in this way. It was found that the agreement between simulation and experiment was not good. This was true also even when other deactivating functions, such as the exponential and hyperbolic forms, were used [equations (7.26) and (7.27)]. It would seem therefore that the use of this method of analysing data on coking may not be of universal application and may break down when complex cracking reactions are considered. In these latter cases, many reactions may be occurring simultaneously, so perhaps it is not altogether surprising that universal agreement is not obtained.

7.3.2 Non-isothermal analysis of coking in fixed bed reactors

The isothermal treatments of coking behaviour in fixed bed reactors can provide information on the general characteristics of coking in many systems of interest. This is, of course, particularly true if the reaction enthalpy is moderate or small, as in many of the isomerization reactions that have been considered in the literature and to which these models have been applied. However, certain phenomena, including the dynamics of fixed beds subject to deactivation, can only be fully analysed using a non-isothermal treatment. Inevitably, a non-isothermal analysis complicates the system because of the necessary introduction of additional parameters. The most important of these are the activation energies for the main and deactivation reactions since the relative magnitude of these will determine whether fouling increases or decreases with changes of temperature within the reactor, but other parameters such as the heats of adsorption for the various species can become significant if Langmuir–Hinshelwood kinetics are appropriate for the system. The necessity for the introduction of the various parameters will be referred to at appropriate points in the text when these arise.

Ervin and Luss (1970) made a very interesting study of non-isothermal fouling in fixed bed reactors. The basic aim of their work was to determine the stability of fixed bed reactors under conditions where fouling was occurring, and this aspect will be referred to later in this chapter. However, profiles of temperature and activity were also obtained under conditions where there was no possibility of reactor instabilities.

An unusual feature of the analysis was that a cell model was used to simulate the reactor. This type of model, first proposed by Deans and Lapidus (1960), assumes that the bed is composed of a series of cells, each well mixed, which are connected by the interstitial fluid. Within each cell account can be taken of both solid and fluid phases, so the heterogeneous nature of the system can be retained. In their application of this model, Ervin and Luss assumed that there were no transport resistances of heat or mass within the catalyst pellet, these resistances being lumped in the gas film. The main reaction A → B was assumed to be of first order and irreversible and its rate is described by:

$$r(B) = k\psi p_p \tag{7.29}$$

where ψ is the normalized activity factor or function and p_p the partial pressure of reactant A in the catalyst. Deactivation by parallel or series fouling was assumed and the appropriate deactivation expressions for these are written as:

$$\frac{d\psi}{d\theta} = -k_p \psi p_p \quad \text{(Parallel fouling, A → C)} \tag{7.30}$$

$$\frac{d\psi}{d\theta} = -k_p \psi p_{pB} \quad \text{(Series fouling, B → C)} \tag{7.31}$$

where θ is the time, k_p the rate constant of the fouling reaction, and p_{pB} the partial pressure of reactant B in the catalyst.

The mass and heat conservation equations for the jth cell are written as:

$$a_1 \frac{dp_j}{d\theta} = \frac{G}{P\bar{m}k_c a_v \delta d_p}(p_{j-1} - p_j) - (p_j - p_{pj}) \tag{7.32}$$

$$a_2 \frac{dT_j}{d\theta} = \frac{Gc_f}{h_f a_v \delta d_p}(T_{j-1} - T_j) - (T_j - T_{pj}) \tag{7.33}$$

$$a_3 \frac{dp_{pj}}{d\theta} = p_j - p_{pj} - k_j \psi_j p_{pj} \tag{7.34}$$

$$a_4 \frac{dT_{pj}}{d\theta} = T_j - T_{pj} + \beta k_j \psi_j p_{pj} \tag{7.35}$$

where

$$k_j = \frac{d_p}{6} \frac{\rho_s S_g k'_j}{k_c} \tag{7.36a}$$

$$\beta = \frac{k_g}{h_f}(-\Delta H) \qquad (7.36b)$$

$$a_1 = \frac{\varepsilon \rho_f}{k_g a_v \bar{m}}; \qquad a_2 = \frac{\varepsilon \rho_f c_f}{a_v h_f} \qquad (7.36c)$$

$$a_3 = \frac{d_p \alpha \rho_f}{6 \bar{m} k_g}; \qquad a_4 = \frac{d_p \rho_s c_f}{6 h_f} \qquad (7.36d)$$

in which $\bar{m}$ is the average molecular weight of the gas mixture, k_g and h_f the film mass and heat transfer coefficients respectively, d_p the particle diameter, S_g the specific surface of the catalyst, a_v the superficial area of the particles per unit bed volume, c_f and c_s the specific heats of gas and solid, and ρ_f and ρ_s the densities of gas and solid respectively. The bed voidage is ε while that of the particles is α.

The above equations are subject to the inlet conditions:

$$T_e = T_e(\theta) \qquad (7.37)$$

$$p_e = p_e(\theta) \qquad (7.38)$$

and the initial conditions:

$$\left.\begin{array}{l} p_j = p_{jo} \\ T_j = T_{jo} \\ p_{pj} = p_{pjo} \\ T_{pj} = T_{pjo} \\ \psi_j = 1 \end{array}\right\} \theta = 0 \quad j = 1, 2, \ldots, n \qquad (7.39)$$

Assuming a pseudo-steady state, since the rate of fouling is usually slow compared with the other processes occurring in the system, equations for all j are:

$$M(p_{j-1} - p_j) = p_j - p_{pj} \qquad (7.40)$$

$$H(T_{j-1} - T_j) = T_j - T_{pj} \qquad (7.41)$$

$$p_j - p_{pj} = k_j \psi_j p_{pj} \qquad (7.42)$$

$$T_{pj} - T_j = \beta k_j \psi_j p_{pj} \qquad (7.43)$$

where M and H are the dimensionless factors multiplying the first bracketed term on the right-hand sides of equations (7.32) and (7.33). In (7.42) and (7.43) k_j is written in the form:

$$k_j = k_0 \exp(-E/RT_{pj}) \qquad (7.44)$$

7 Deactivation in Catalytic Reactors

Table 7.1 Parameters used in the simulation of Ervin and Luss (1970)

$H = 2.352$	$M = 3.919$
$\beta = 6000°\text{F min}^{-1}$	$a_4 = 1/5.45 \text{ min}$
$k_j = \exp(12.98 - 22\,000/T_{pj})$	$k_p = A_p \exp(-10\,000/T_{pj})$

The transient behaviour of the bed was simulated by Ervin and Luss using 80 cells. The frequency factor of the fouling rate constant was assigned a value that caused the reactor to be fouled after 1350 min on stream. The values of the dimensionless groups chosen are listed in Table 7.1. Thus the activation energy of the fouling process was taken as smaller than that for the main reaction.

The transient behaviour of the reactor was simulated from equations (7.30)–(7.36) for two types of steady state analysis. The pseudo-steady state equations (7.40)–(7.43) were used to predict whether a unique pseudo-steady state would be obtained or non-unique steady states would arise. Since the latter is in the province of reactor stability only the unique pseudo-steady state case will be considered in this section. Figure 7.9 shows the transient temperature profiles for parallel fouling for the unique pseudo-steady state case. For this case the entry partial pressure of reactant was 0.07 and the pre-exponential factor for fouling was taken as 25.0. It should be noted that the form of the deactivation relations (7.30) and (7.31) is equivalent to the exponential activity function used by Froment and Bischoff. Figure 7.9 demonstrates that the reactor would operate at constant conversion for a long period of time but that the reaction zone is continuously moving downstream. Eventually, at 1350 min, the reaction zone moves out of the

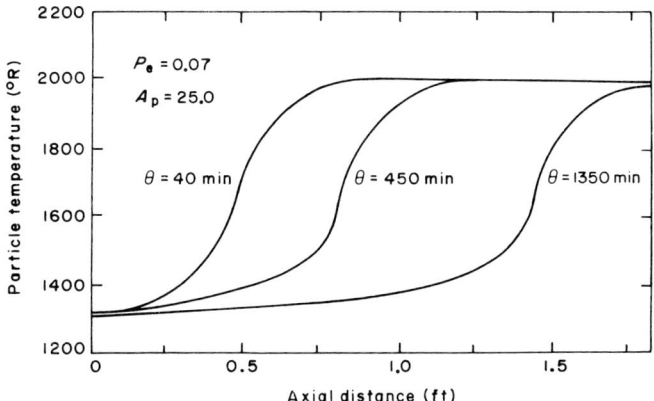

Fig. 7.9 Transient temperature profiles for fouling by a parallel reaction for a case where each particle has a unique pseudo-steady state. (Ervin and Luss, 1970)

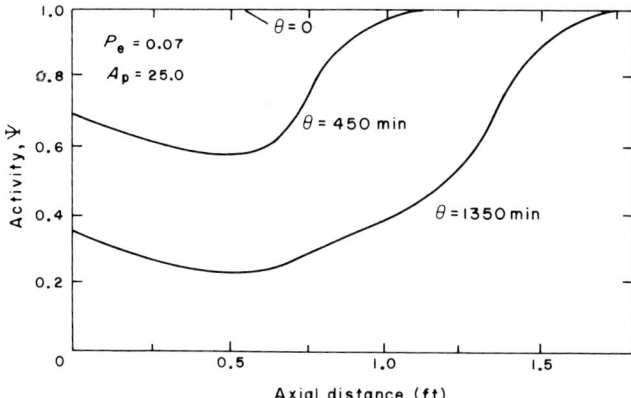

Fig. 7.10 Activity profiles for the case shown in Fig. 7.9. (Ervin and Luss, 1970)

reactor and a sudden decrease in conversion occurs. The corresponding activity profiles shown in Fig. 7.10 indicate that most of the deactivation occurs in the same region as the original reaction zone. Although the activity profiles flatten with time, the exit of the bed is still practically unpoisoned even after 1350 min.

The situation when series fouling occurs is illustrated in Fig. 7.11. The temperature profile ahead of the reaction zone is now continuously moving downwards and tends to "recline" with process time. At 1350 min the temperature profile is almost linear while the conversion has dropped to 80%

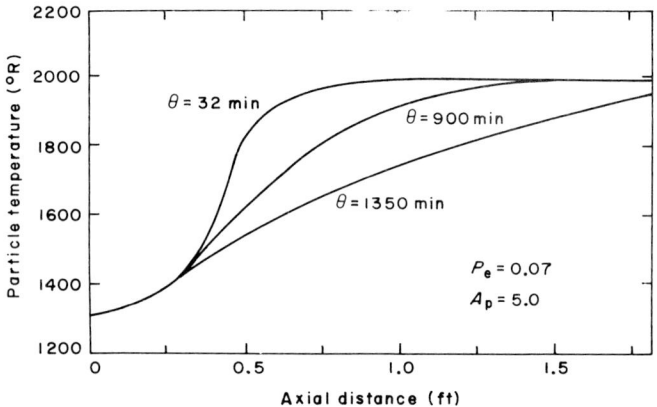

Fig. 7.11 Transient temperature profiles for fouling by a consecutive reaction for a case where each particle has a unique pseudo-steady state. (Ervin and Luss, 1970)

7 Deactivation in Catalytic Reactors

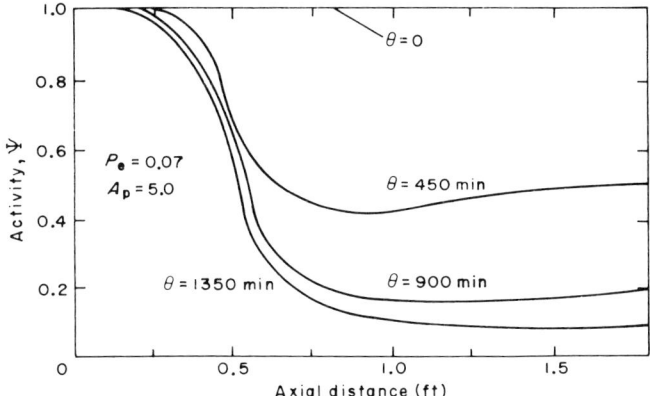

Fig. 7.12 Activity profiles for the case shown in Fig. 7.11. (Ervin and Luss, 1970)

of the initial value. The corresponding activity profile is shown in Fig. 7.12; as expected from previous discussion, the deactivation is most prominent in the region after the reaction zone where the temperature and concentration of B are high. The rate of fouling also decreases with time owing to deactivation of the pellets.

These results are important in that a full transient analysis was employed and the resulting activity profiles were obtained under non-isothermal conditions. However, simple first order kinetics were assumed for the main reaction and intraparticle gradients of temperature and concentration were neglected. Intraparticle temperature gradients may well be insignificant for a reactive catalyst but concentration gradients within the catalyst cannot generally be ignored.

Kam and Hughes (1979b) developed a model for adiabatic fixed bed reactors which included intraparticle gradients of temperature and concentration. Additionally, Langmuir–Hinshelwood kinetics were adopted for the main reaction although the fouling reaction was considered to be of first order. For this analysis it was desired to focus attention solely on the deactivating effects of fouling by coke deposition, and therefore the complicating features of multiplicity of steady states were avoided by parameter selection. Therefore steep temperature and concentration gradients did not appear in the bed profiles, and in many cases the reactant concentration was finite and not zero at the exit of the reactor. The treatment used in modelling was based on that used in the treatment of single particles subject to coking as developed by Kam et al. (1977) and Kam and Hughes (1979a). It was shown that the activation energy parameter for fouling, γ_f (analogous to the Arrhenius number γ for the main reaction), was an important parameter and, as expected with adiabatic operation for an exothermic main reaction, more deactivation occurred as γ_f was increased.

The disadvantage of this approach is a conceptual one. Since the coking reactions are necessarily linked to the main reaction by processes either in parallel or in series, it would seem logical to postulate that, if the main reaction occurs by a Langmuir–Hinshelwood mechanism, the fouling reaction should proceed via a similar mechanism. Dumez and Froment (1976) have used a Langmuir–Hinshelwood form for the coking reaction and coupled this with an empirical activity function to determine the coking rate. However, their results were confined to one endothermic reaction, namely dehydrogenation of butene to butadiene, and no exothermic reactions were considered. The analysis of Kam and Hughes (1979b) was extended therefore to include the case where the coking occurred via a Langmuir–Hinshelwood mechanism (Brito-Alayon et al., 1981, 1982). The assumptions made in the analysis were: (1) plug-flow operation; (2) the fouling reaction occurs much more slowly than the main reaction, so pseudo-steady state conditions apply; (3) the gas stream has a constant average heat capacity and density; (4) the heat capacity of the reactor wall is negligible; (5) the effect of volume change due to reaction is neglected; (6) the external heat and mass transfer coefficients are not influenced by surface reaction; (7) particle–particle heat conduction and heat transfer by radiation are negligible; (8) the particle effective diffusivity and thermal conductivity remain constant.

For a heterogeneous model the heat and mass balance inside the particles as well as in the gas streams have to be considered. On the other hand, fouling occurs only on and within the catalyst pellets, and no deactivation equation is necessary in the external field formulation.

External field heat and mass balances

With the assumptions stated above, the pseudo-steady state mass and energy balance equations for the adiabatic bed can be written as follows:

$$\varepsilon' u \frac{dC_{Ao}}{dz} = -a_v k_c (C_{Ao} - C_{As}) \qquad (7.45)$$

$$\varepsilon' u \rho_g c_{pg} \frac{dT_o}{dz} = a_v h (T_s - T_o) \qquad (7.46)$$

where k_c and h are the film mass and heat transfer coefficients and a_v the external surface area of the pellets per unit bed volume.

By taking a heat and mass balance around the catalyst particles, the concentration and temperature flux terms can be expressed as:

$$-\frac{a_v k_m}{\varepsilon'}(C_{Ao} - C_{As}) = (1 - \varepsilon') r_A(C_{Ao}, T_o) \times \eta(C_{Ao}, T_o, \bar{S}) \qquad (7.47)$$

$$-\frac{a_v h}{\varepsilon'}(T_s - T_o) = (1 - \varepsilon')(-\Delta H)r_A(C_{Ao}, T_o) \times \eta(C_{Ao}, T_o, \bar{S}) \quad (7.48)$$

Thus equations (7.45) and (7.46) become:

$$\frac{dC_{Ao}}{dz} = \frac{(1-\varepsilon')}{u} r_A(C_{Ao}, T_o) \times \eta(C_{Ao}, T_o, \bar{S}) \quad (7.49)$$

$$\frac{dT_o}{dz} = -\frac{(1-\varepsilon')(-\Delta H)}{u\rho_g c_{pg}} r_A(C_{Ao}, T_o) \times \eta(C_{Ao}, T_o, \bar{S}) \quad (7.50)$$

In these equations $\bar{S}$ is the mean catalyst activity.

The boundary conditions are:

$$z = 0 \quad C_{Ao} = C_{Ao}|_{z=0} \quad (7.51a)$$

$$T_o = T_o|_{z=0} \quad (7.51b)$$

These equations can be put into dimensionless form by introducing the following variables:

$$\xi = z/L \quad (7.52a)$$

$$a_o = \frac{C_{Ao}}{C_{Ao}|_{z=0}} \quad (7.52b)$$

$$\theta = \frac{T_o}{T_o|_{z=0}} \quad (7.52c)$$

$$\Omega = \frac{k(1-\varepsilon')L}{u}(C_{Ao}|_{z=0})^{n-1} \quad (7.52d)$$

$$\beta' = \frac{-\Delta H}{\rho_g c_{pg}}\left(\frac{C_{Ao}}{T_0}\right)\bigg|_{z=0} \quad (7.52e)$$

where ξ is the dimensionless coordinate along the packed bed, a_o and θ_o are the dimensionless bulk concentration and temperature, n is the order of reaction, and Ω and β' are the reaction modulus and thermicity factor in the external field, respectively. The reaction modulus Ω is a modified form of the Damköhler number which, as defined, represents the ratio of the reaction rate to the linear flow rate of the reactants.

The dimensionless forms of equations (7.49) and (7.50) become:

$$\frac{da_o}{d\xi} = \Omega f(a_o, \theta_o)\eta(a_o, \theta_o, \bar{S}) \quad (7.53)$$

$$\frac{d\theta_o}{d\xi} = -\beta'\Omega f(a_o, \theta_o)\eta(a_o, \theta_o, \bar{S}) \quad (7.54)$$

and the dimensionless boundary conditions are:

$$\xi = 0 \quad a_o = 1 \tag{7.55a}$$
$$\theta_o = 1 \tag{7.55b}$$

The particle effectiveness factor, $\eta(a_o, \theta_o, \bar{S})$, is obtained by solving the mass and energy continuity equations and may be expressed as:

$$\eta(a_o, \theta_o, \bar{S}) = \frac{\alpha \int_0^1 \delta^{(\alpha-1)} f(a, \theta) S \, d\theta}{f(a_o, \theta_o)} \tag{7.56}$$

where S and $\bar{S}$ are the point and mean activity of the pellet and α is the geometric shape factor for the pellets.

Single-particle heat and mass balance

For the case of non-isothermal fouling, and with the inclusion of external film resistances, the dimensionless equations are:

$$\frac{d^2 a}{d\delta^2} + \frac{\alpha - 1}{\delta} \frac{da}{d\delta} = \phi^2 f(a) S \exp\left[\gamma\left(1 - \frac{1}{\theta}\right)\right] \tag{7.57}$$

$$\frac{d^2 \theta}{d\delta^2} + \frac{\alpha - 1}{\delta} \frac{d\theta}{d\delta} = -\beta \phi^2 f(a) S \exp\left[\gamma\left(1 - \frac{1}{\theta}\right)\right] \tag{7.58}$$

$$\tau = 0; \quad S = 1 \text{ for any } \delta \tag{7.59}$$

$$\delta = 0; \quad \frac{da}{d\delta} = \frac{d\theta}{d\delta} = 0 \text{ for any } \tau \tag{7.60}$$

$$\delta = 1; \quad \frac{1}{Sh^*} \frac{da_s}{d\delta} = (a_o - a_s) \tag{7.61}$$

$$\frac{1}{Nu^*} \frac{d\theta_s}{d\delta} = (\theta_o - \theta_s) \tag{7.62}$$

where δ is the dimensionless particle radius and γ the Arrhenius number for the main reaction, and where use is made of the following stoichiometric relation between the concentration of reactant A and product B assuming equimolar counterdiffusion:

$$b = b_o + q(a_o - a) \tag{7.63}$$

The temperature may be expressed in terms of concentration via the Prater relation:

$$\theta = \theta_o + \frac{Sh^*}{Nu^*} \beta(a_o - a_s) + \beta(a_s - a) \tag{7.64}$$

7 Deactivation in Catalytic Reactors

The reaction kinetics for the main reaction are taken to be of Langmuir–Hinshelwood form and may be written as:

$$f(a, \theta) = \frac{a^n \exp[(\gamma - h_{K_A})(1 - 1/\theta)]}{1 + K_A^* a \exp[-h_{K_A}(1 - 1/\theta)] + K_B^*(a_o + b_o - a)\exp[-h_{K_B}(1 - 1/\theta)]} \quad (7.65)$$

The dimensionless rate equation for the deactivation can be assumed to be of the generalized form:

$$\frac{dS}{d\tau} = g_1(a, \theta)S + \frac{k_{f2}}{k_{f1}} g_2(b, \theta)S \quad (7.66)$$

where $g_1(a, \theta)$ and $g_2(a, \theta)$ are functions of concentration and temperature and may be expressed in the form of simple integral kinetics or in complex kinetic form. The term k_{f1} signifies parallel fouling while k_{f2} is that for series fouling. In the work described two different types of fouling kinetics were studied. In the first case, fouling was assumed to be first order in the concentration of fouling precursor and also first order in active sites concentration:

$$g_1(a, \theta) = a \exp[\gamma_f(1 - 1/\theta)] \quad (7.67)$$

$$g_2(b, \theta) = b \exp[\gamma_f(1 - 1/\theta)] \quad (7.68)$$

When Langmuir–Hinshelwood kinetics are assumed for the deactivation reaction, fouling was assumed to be a function of the active sites concentration and the complex kinetics described by g_1 and g_2 above. The functions g_1 and g_2 are now written as:

$$g_1(a, \theta) = \frac{K_A^* a \exp[(\gamma_f - h_{K_A})(1 - 1/\theta)]}{1 + K_A^* a \exp[-h_{K_A}(1 - 1/\theta)]} \quad (7.69)$$

$$g_2(b, \theta) = \frac{K_B^* b \exp[(\gamma_f - h_{K_B})(1 - 1/\theta)]}{1 + K_B^* b \exp[-h_{K_B}(1 - 1/\theta)]} \quad (7.70)$$

The appropriate equations for external field and pellet were solved numerically.

A comparison was made between this model and fouling by a first order mechanism and results were obtained for the parameter values given in Table 7.2.

Table 7.2 Values of parameters.

Sh*	Nu*	K_A^*, K_B^*	h_{K_A}, h_{K_B}	γ	γ_f	Ω
250	1.5	10	−5	20	40	16

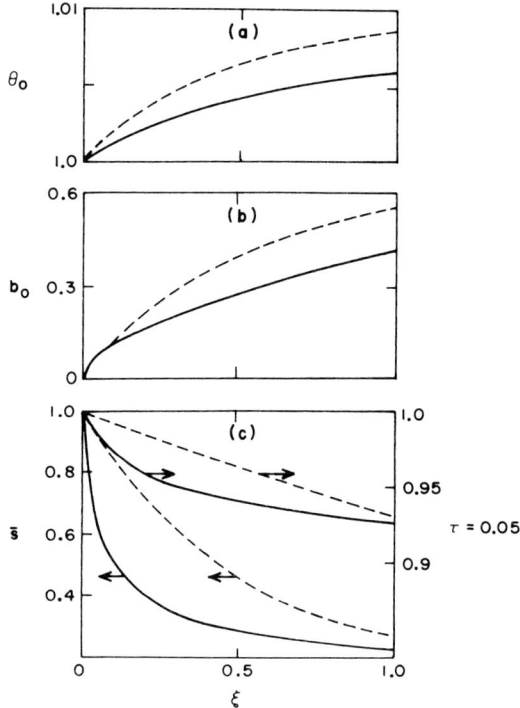

Fig. 7.13 Dimensionless temperature, product concentration, and activity profiles for exothermic reaction with series fouling: (———) Langmuir–Hinshelwood fouling; (– – – –) first order fouling. (Brito-Alayon, 1981)

The results obtained for an exothermic reaction with fouling by the product are shown in Fig. 7.13. The temperature rise in the reactor, the product concentration, and the activity in the reactor are each plotted as a function of dimensionless reactor length for a value of the dimensionless time τ equal to 1.2. The values of the thermicity factors employed are $\beta = 0.02$ for the pellet and $\beta' = 0.015$ for the packed bed, while the activation energy parameter for fouling, γ_f, is taken as 40 for both first order and Langmuir–Hinshelwood fouling. The essential feature of Fig. 7.13 (a) and (b) is that more reaction occurs with first order fouling compared with Langmuir–Hinshelwood fouling due to the higher effective reaction order in the former case. This is demonstrated by the larger increase in temperature and the increased product concentration. It is also observed that there is no significant difference in product concentration over the first 10% of the reactor length for either type of fouling.

7 Deactivation in Catalytic Reactors

The corresponding activity profiles for the two types of fouling at $\tau = 1.2$ and 0.05 are shown in Fig. 7.13(c). For both first order and Langmuir–Hinshelwood fouling, the activity is highest at the entrance to the reactor and decreases with distance along the reactor. This is in accord with previous findings for a series fouling mechanism, since the fouling precursor is the product and the concentration of this will be zero at the reactor entrance (for no recycle stream) and will progressively increase along the reactor. However, an important difference between the two types of fouling is evident; Langmuir–Hinshelwood fouling gives a sharp drop in activity close to the reactor entrance, whereas for first order fouling the activity decreases more gradually from inlet to exit.

One explanation for the initial sharp decrease in catalyst activity for Langmuir–Hinshelwood fouling at the reactor entrance is attributable to the difference in mechanism between this and first order fouling. At the reactor entrance the main reaction will give only a small amount of product. For Langmuir–Hinshelwood fouling this product can give an immediate coke deposit but for first order fouling this product is present in the gas space in the pores of the pellet and may not be immediately available for depositing coke.

It is interesting to note that the initial decrease in activity for Langmuir–Hinshelwood fouling in Fig. 7.13(c) is very pronounced. Examination of Fig. 7.13(a) reveals that, although Langmuir–Hinshelwood fouling gives a smaller temperature rise, both types of fouling show small increases in reactor temperature which are relatively close. Because of this relatively small difference in reactor temperature, comparison of equations (7.68) and (7.70) shows that Langmuir–Hinshelwood deactivation is greater than that for first order by a factor approximating to $K_B^*(1 + K_B^* b)$, provided that the amount of fouling precursor is the same for both processes. This requirement is fulfilled for the first 10% of reactor length [Fig. 7.13(b)] and results in greater deactivation by a Langmuir–Hinshelwood mechanism in this region [Fig. 7.13(c)]. Thereafter, the precursor concentration is less than that for first order fouling, so the activity profile levels off with increase in reactor length [Fig. 7.13(c)]. In fact, the activity for Langmuir–Hinshelwood fouling under these conditions is always less than that for first order fouling, as examination of the activity profiles at a reaction time of $\tau = 0.05$ shows [upper part of Fig. 7.13(c)]. For this short process time the precursor (product) concentration and temperature profiles for first order and Langmuir–Hinshelwood mechanisms are almost identical, but Langmuir–Hinshelwood fouling gives a much more rapid activity decrease, particularly at the front of the reactor.

A comparison of the temperature, reactant concentration, and activity profiles for parallel fouling under exothermic conditions is given in Fig.

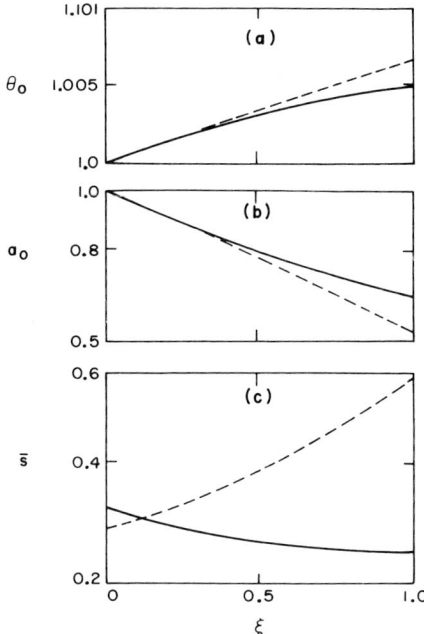

Fig. 7.14 Dimensionless temperature, reactant concentration, and activity profiles for exothermic reaction with parallel fouling: (———) Langmuir–Hinshelwood fouling; (– – – –) first order fouling. (Brito-Alayon *et al.*, 1981)

7.14(a–c). The most surprising effect shown by the activity profiles [Fig. 7.14(c)] is that Langmuir–Hinshelwood fouling for these parameter values shows behaviour completely contradictory to what may be termed "normal" fouling behaviour for fouling by a parallel mechanism. "Normal" behaviour is shown by the first order fouling activity profile, in which the activity is least at the reactor entrance and increases with distance along the reactor. This behaviour is expected, since the fouling precursor (in this case the reactant) has the greatest concentration at the entrance to the reactor. Therefore fouling would be expected to be predominant at the entrance and to decrease thereafter as the reactant concentration diminished along the length of the reactor. The Langmuir–Hinshelwood activity profile in Fig. 7.14(c), however, shows maximum catalytic activity at the reactor entrance, with a further slight but steady decline with distance along the reactor. Such a profile would in general be recognized as being more typical of series fouling (except that in this case there would be no initial low activity level at the reactor entrance). Furthermore, the overall activity of the bed is much less than that for first order fouling.

This reversal of the normal parallel fouling activity pattern in a fixed bed reactor has important consequences if the effect is general for Langmuir–Hinshelwood fouling, since a large number of gas–solid reactions obey Langmuir–Hinshelwood kinetics.

The effect of diffusional resistance on activity profiles when both the main and coking reactions obey Langmuir–Hinshelwood kinetics was also determined by Brito-Alayon et al. (1982). The simulated activity profiles were obtained as three-dimensional plots in which the vertical axis represents the mean activity of the pellet at that point in the reactor, while one horizontal axis gives values of the Thiele modulus, ϕ, plotted on a logarithmic scale. Values of the Thiele modulus ranged from 2.5 to 30 and γ_f was taken as 40. In all figures the fouling time, τ, was equal to 1.2, corresponding to a time on stream of 18 days for the parameters considered.

Figures 7.15(a) and 7.15(b) show the effect of increasing Thiele modulus on the activity profiles for endothermic reactions with $\beta = -0.015$ for the pellet and $\beta' = -0.020$ for the reactor. For the case of series fouling Fig. 7.15(a) shows that the reactor activity is always greatest at the inlet of the reactor for the whole range of ϕ investigated ($\phi = 2.5\text{--}30$). At the reactor outlet the activity was least and almost independent of ϕ at an activity level of about 0.55.

The decrease in activity from inlet to outlet is greatest for low values of ϕ and reflects the pronounced effect of Thiele modulus on activity at the reactor inlet. The results show the same pattern as observed previously on single catalyst pellets obeying first order kinetics for coking (Kam et al., 1975, 1977b). This is to be expected since conditions at the inlet of a plug flow reactor operating without recycle would be virtually the same as for a single pellet. The effect of an increase in ϕ is to increase the product concentration inside the pellet. Since this is the precursor of fouling by a series mechanism, the extent of deactivation would be expected to decrease as ϕ increases, as was observed.

In contrast, when parallel fouling of an endothermic main reaction is considered, the situation is much more complicated, as shown in Fig. 7.15(b). At low values of ϕ (between 2.5 and 10) the activity of the reactor is least at $\xi = 0$ and gradually increases throughout the reactor. It should be noted that the activity is very low at the inlet at about 0.4 and is less than 0.6 at the outlet. This behaviour is consistent with the usually accepted distribution of activity when parallel fouling occurs, since now the reactant is the fouling precursor and its concentration is greatest at the reactor inlet. At values of the Thiele modulus greater than 10, however, the activity profile is reversed, with the activity now greatest at the reactor inlet and decreasing progressively along the length of the reactor. An explanation for this effect can be given in terms of the distribution of fouling deposit at low and high values of

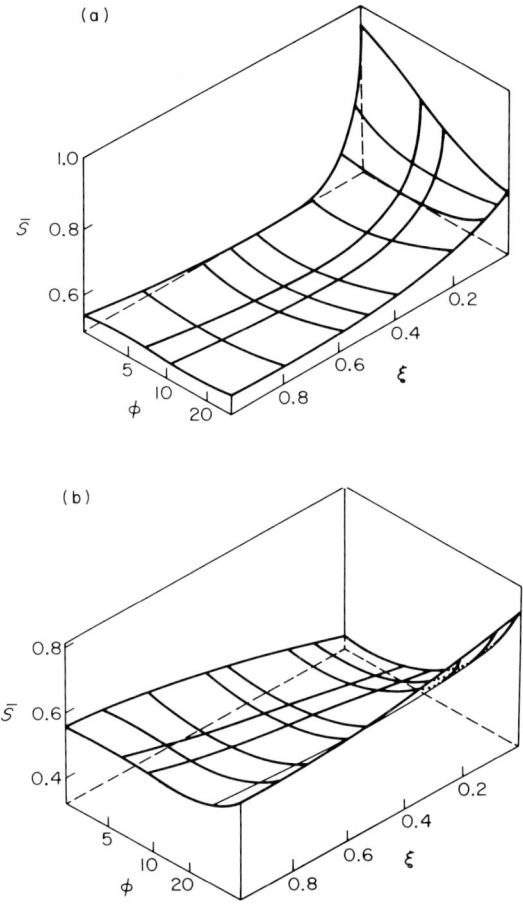

Fig. 7.15 Variation of activity profiles with Thiele modulus for an endothermic reaction with (a) series fouling and (b) parallel fouling: $K_A^* = K_B^* = 10$. (Brito-Alayon et al., 1982)

the Thiele modulus. At low values of ϕ the reactant is distributed uniformly throughout the pellet and therefore deposition of foulant will also occur throughout the pellet. At higher values of the Thiele modulus, reaction is confined to a relatively thin zone near the surface of the pellet, due to the diffusional restriction. Under such conditions it has been shown that the mean activity of the pellet may be greater than for the uniform deactivation obtained with a low value of ϕ (Kam et al., 1977b). Another factor contributing to the unexpected activity distribution at $\xi = 0$ is the non-isothermal nature of the reaction. At low values of ϕ the reaction rate is relatively slow and hence the effect of temperature is not significant. When ϕ is large, however, the rate of the main reaction is high and this leads to a greater temperature reduction due to the reaction endothermicity which will favour the main reaction for which γ is 20 while causing a relative decrease in the rate of the deactivation reaction with the larger γ_f value of 40.

At the reactor exit, Fig. 7.15(b) shows that there is only a small difference in activity for the whole range of ϕ. This is due to the decrease in temperature caused by the endothermic nature of the reaction reducing the rate of both main and deactivation reactions in this region. It is of interest to note that, at high values of ϕ for this case of endothermic parallel fouling, the mean activity of the reactor is greater than that at low values of ϕ. Thus it may be preferable to operate under diffusion controlled conditions when γ_f is greater than γ as in this instance. Furthermore, a comparison of Figs 7.15(a) and 7.15(b) shows that the value of ϕ selected affects the deactivation behaviour of the reactor more at the inlet than at the exit for both fouling mechanisms.

When the main reaction is exothermic the results obtained for deactivation by a series of fouling mechanisms are shown in Fig. 7.16(a). The activity profiles obtained are generally similar to those shown in Fig. 7.15(a) for endothermic series fouling, but now there is a much lower level of activity of about 0.2 at the reactor exit for all values of ϕ. Also at the inlet of the reactor ($\xi = 0$), a much more increased rate of deactivation occurs as the value of ϕ is increased compared with the endothermic reaction. As for Fig. 7.15(a), an increase in the value of ϕ will increase the product concentration inside the pellet, and for series fouling this will lead to increased deactivation. However, for endothermic conditions this effect was reduced because of the endothermic temperature decrease which becomes greater as the value of ϕ increases. In the present exothermic case, however, the temperature increase in the pellets as ϕ increases will increase the rate of deactivation, since γ_f is greater than γ and will thus give a much more severe decrease in activity at the reactor inlet as ϕ is increased; Fig. 7.16(a). Results for the surface temperature distribution in the first layer of pellets are given in Tables 7.3 and 7.4 for endothermic and exothermic reactions respectively.

Table 7.3 Particle surface temperatures for endothermic reactions at $\xi = 0$.

Parallel fouling $\theta_s\|_{\tau=1.0}$	ϕ	Series fouling $\theta_s\|_{\tau=1.0}$
0.99946	7	0.99890
0.99876	10.5	0.99783
0.99035	30	0.98878
0.97936	50	

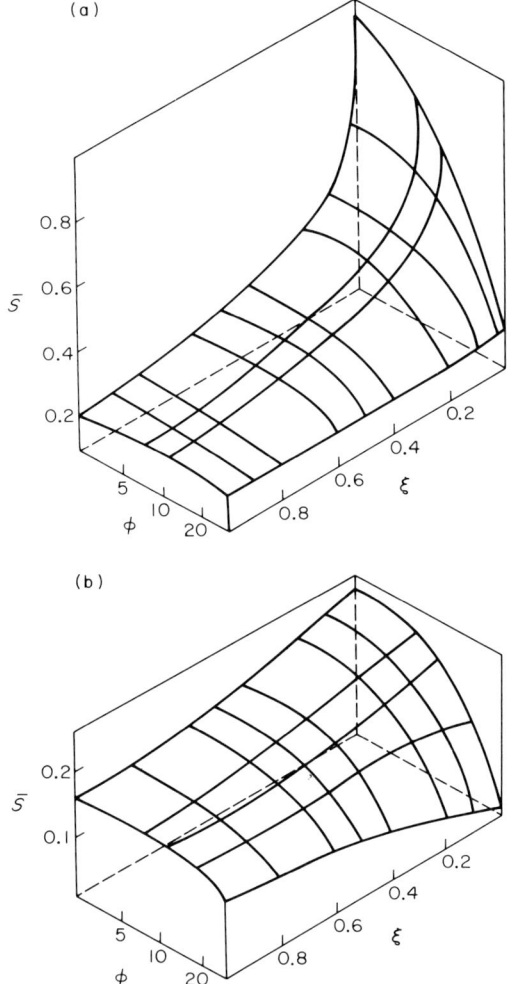

Fig. 7.16 Variation of activity profiles with Thiele modulus for an exothermic reaction with (a) series fouling and (b) parallel fouling; $K_A^* = K_B^* = 10$. (Brito-Alayon et al., 1982)

The increase in temperature along the reactor for the exothermic reaction also accounts for the reduced activity at the exit, since the deactivation reaction is favoured by an increase in temperature.

Figure 7.16(b) shows the activity profiles for parallel fouling under exothermic conditions. At values of ϕ between 2.5 and 10 there is a steady decrease of activity from $\xi = 0$ to $\xi = 1$, although the degree of activation is

7 Deactivation in Catalytic Reactors

Table 7.4 Particle surface temperatures for exothermic reactions at $\xi = 0$

Parallel fouling $\theta_s\vert_{\tau=1.0}$	ϕ	Series fouling $\theta_s\vert_{\tau=1.0}$
1.00009	2.5	1.00022
1.00067	7.0	1.00150
1.00142	10.5	1.00297
1.00475	30.0	1.00962

severe at all points along the reactor. This behaviour is again not typical of the "normal" behaviour associated with parallel fouling. When ϕ is increased above 10, normal type behaviour is observed for parallel fouling, with the activity least at the reactor inlet and increasing along the bed. It was observed previously (Brito-Alayon et al., 1981) that, when first order fouling was used in the simulations at this value of the Thiele modulus, no such anomaly occurred. This effect therefore demonstrates that when Langmuir–Hinshelwood kinetics describe the reactions the diffusional resistance may have a profound influence on the shape of the activity profile.

All the above modelling results were obtained for values of K_A^* and K_B^* both equal to 10. To investigate whether the anomalous profiles observed above might be caused by adsorption, simulations where K_A^* and K_B^* were varied from 1 to 40 were made by Brito-Alayon et al. (1982).

The effect of variations in K_B^*, the adsorption constant for series fouling, is shown in Fig. 7.17 for an exothermic main reaction. Profiles are compared at values of ϕ equal to 2.5, where the main reaction would be under chemical control. This value of ϕ was selected because, for values of ϕ greater than 10, activity profiles along the reactor were generally much less steep than those for lower ϕ values, so the effect of parameter variations would be more difficult to determine. K_B^* was varied from 1 to 40 at a constant value of K_A^* equal to 10. The activity profiles in Fig. 7.17 show a continuous decrease from $\xi = 0$ to $\xi = 1$, from a value close to unity at the inlet to the reactor to about 0.3–0.4 at the outlet. With an increase in K_B^*, the decrease of activity in the inlet region of the reactor shows a much steeper decrease than at low values of K_B^*. This gradient is particularly steep for values of K_B^* greater than 20, and for this range of K_B^* values the major portion of the activity decrease in the reaction occurs in the region $\xi = 0$–0.2. The profiles show the typical characteristics associated with series fouling in that the activity is greatest at $\xi = 0$ and least at $\xi = 1$. It is of interest to note, however, that at $\xi = 0$ there is a gradual decrease in activity with increase in K_B^*. The decrease in activity at $\xi = 0$ is less than that when ϕ is increased. This is because an increase in K_B^* does not affect the distribution of foulant or the amount of precursor

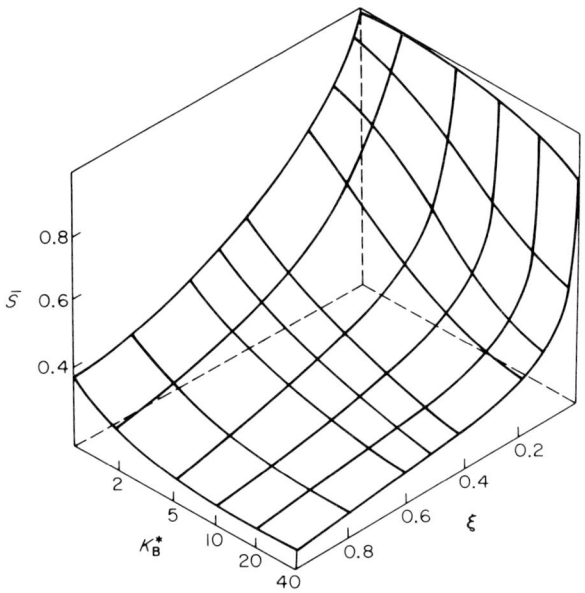

Fig. 7.17 Effect of K_B^* on activity profiles. Exothermic reaction with series fouling; $K_A^* = 10$, $\phi = 2.5$. (Brito-Alayon et al., 1982)

present, only its magnitude. Pellet surface temperatures in the first row in the reactor confirm this, as examination of Table 7.5 shows. These are tabulated for the pellets at the entrance of the reactor for two values of τ equal to 0.05–1.0, respectively. The surface temperatures at $\tau = 0.05$ show only a very slight decrease with increase in the K_B^* values, but at $\tau = 1.0$ a more significant decrease in θ_s is observed showing the effect of deactivation. Deactivation at $\xi = 0$ in this case cannot be attributed to increase in product concentration because ϕ is low; neither can an increase in temperature favouring the fouling process be responsible since, as shown in Table 7.5, in the first pellets

Table 7.5 Particle surface temperature for various values of the adsorption constants of $\xi = 0$.

Parallel fouling		K_A^* or K_B^*	Series fouling	
$\theta_s\|_{\tau=0.5}$	$\theta_s\|_{\tau=1.0}$		$\theta_s\|_{\tau=0.05}$	$\theta_s\|_{\tau=1.0}$
1.00287	1.00141	2	1.00023	1.00023
1.00077	1.00032	5	1.00023	1.00023
1.00000	1.00000	100	1.00022	1.00018

7 Deactivation in Catalytic Reactors

a decrease in temperature is obtained. It can only be concluded that, with large values of K_B^*, adsorption of the small amount of product present is favoured, so this leads to some deactivation even in the first row of pellets for series fouling.

The effect of K_A^* on the activity profiles at a value of ϕ equal to 2.5 is given in Fig. 7.18 for exothermic parallel fouling by a Langmuir–Hinshelwood mechanism. This figure shows that at low values of K_A^* (1 to 5) the "normal" activity characteristics of parallel fouling are observed, i.e. activity is least at the inlet where the reactant concentration is highest. This would be expected because at these low values of K_A^* only a comparatively small amount of reactant will be adsorbed on the active sites of the catalyst, and therefore only small amounts are available for both the main and fouling reactions. This is very similar to the situation for first order fouling kinetics, and therefore the activity profiles would be expected to be similar. At $\xi = 0$ there is a fairly steep decrease in activity as K_A^* is increased. Table 7.5 confirms that, at both short and relatively long fouling times ($\tau = 0.05$ and 1.0 respectively), there is a decrease in surface temperature, when parallel fouling is present, as K_A^* is increased. The decrease in surface temperature is greater, however, for larger values of τ. Therefore the turnover in activity profiles from a rising profile throughout the reactor to a falling one at K_A^* values greater than 5 must be attributed to increased adsorption of reactant, the fouling precursor, in the first row of catalyst pellets, leading to increased deposition of coke.

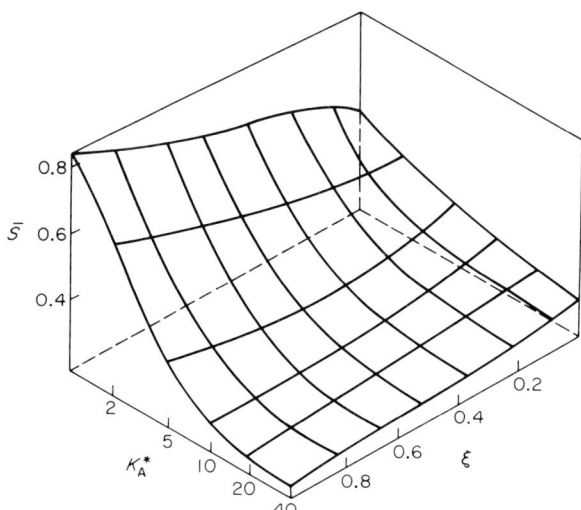

Fig. 7.18 Effect of K_A^* on activity profiles. Exothermic reaction with parallel fouling; $K_B^* = 10$, $\phi = 2.5$. (Brito-Alayon et al., 1982)

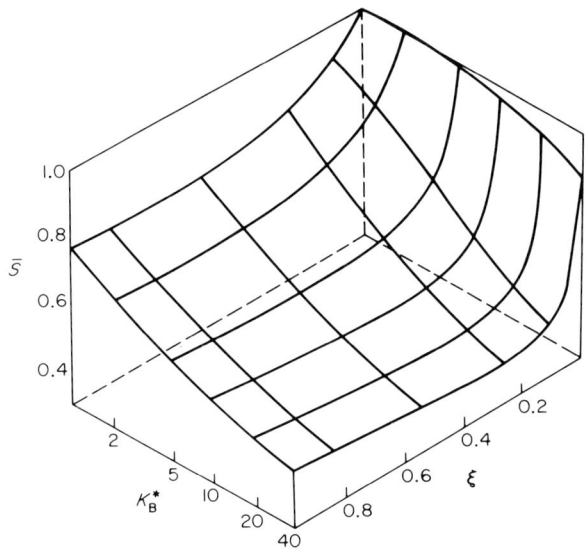

Fig. 7.19 Effect of K_B^* on activity profiles. Endothermic reaction with series fouling; $K_A^* = 10$, $\phi = 2.5$. (Brito-Alayon et al., 1982)

These results are entirely consistent with the single pellet simulations made previously (Kam and Hughes, 1979b).

The influence of the adsorption constants K_A^* and K_B^* for endothermic reactions is given in Figs 7.19 and 7.20 for series and parallel fouling respectively. Figure 7.19 depicts the variation of activity profiles with K_B^* for series fouling. As can be seen, for all values of K_B^* the activity profiles fall monotonically from the entrance to the exit of the reactor. Comparison with Fig. 7.17 for exothermic reaction shows that the endothermic reaction gives less deactivation for all values of K_B^*. This is to be expected because of the high value of γ_f (40). However, even under these milder endothermic conditions, increase in K_B^* causes a sharp decrease in the activity profiles. This is especially pronounced at $K_B^* = 40$ where the activity drops from 0.79 to 0.4 within the first 10% of the reactor length.

Figure 7.20 shows the corresponding activity profiles for parallel fouling with an endothermic main reaction. In this case, at low values of the adsorption constant K_A^*, the behaviour is "normal" in that the activity increases from inlet to exit of the reactor. This behaviour is perpetuated for all values of K_A^*, although when K_A^* is greater than about 20 the activity profiles are almost flat throughout the whole reactor length at an activity level close to 0.3.

All these results demonstrate that the diffusional resistance and adsorption

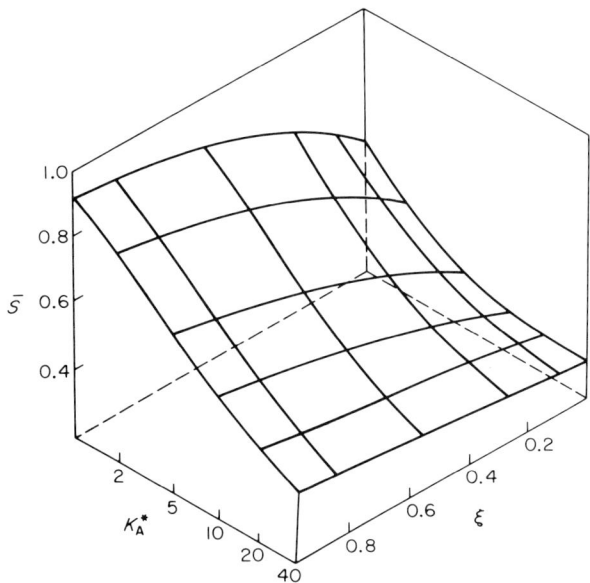

Fig. 7.20 Effect of K_A^* on activity profiles. Endothermic reaction with parallel fouling; $K_B^* = 10$, $\phi = 2.3$. (Brito-Alayon et al., 1982)

constants may seriously modify the accepted pattern of activity profiles characterizing series and parallel fouling. The effect is particularly important for parallel fouling where the "normal" activity profile may be reversed. This anomalous behaviour occurs only for Langmuir–Hinshelwood rate expressions but, since a large number of important reactions are believed to proceed via such mechanisms, care should be exercised in making predictions of activity profiles for such reactions throughout the whole range of Thiele modulus and adsorption constants.

7.4 Thermal sintering of reactors

In any deactivation process involving catalytic reactors it is frequently difficult to assign any particular mode of deactivation to a given process. Thus, poisoning may occur simultaneously with coking of a catalyst, and either of these may be associated individually or collectively with the sintering of a catalyst. An example of simultaneous coking and poisoning due to sulphur species occurs with high nickel content catalysts used in the steam reforming of methane (Moseley et al., 1972). Another example is the bifunctional platinum on alumina catalysts used in naphtha reforming where

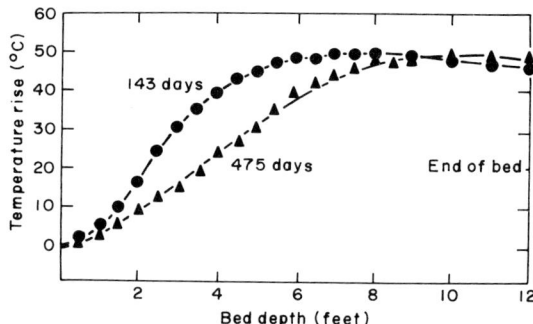

Fig. 7.21 Two temperature profiles obtained at different indicated times, but with similar operating conditions, for a catalyst losing activity by thermal sintering. (Denny and Twigg, 1980)

coking can be appreciable but is often accompanied by sintering of the platinum crystallites.

One example where the deactivation appears to be clearly attributable to thermal sintering alone is that quoted by Denny and Twigg (1980) for a high temperature shift catalyst. The profiles obtained for two times on stream are shown in Fig. 7.21 and indicate that deactivation is occurring throughout the bed, due to this slow thermal sintering. The catalyst in this case was an Fe_2O_3–Cr_2O_3 high temperature shift catalyst, and such behaviour would be expected owing to the nature of the sintering phenomena.

Other examples of thermal sintering occur with silica–alumina catalysts when these are regenerated. If care is not exercised over the combustion conditions, temperatures within the catalyst can rise to very high levels with consequent collapse of the pore structure and loss of active area.

7.5 Reactor dynamics and catalyst deactivation

The question arises whether deactivation of a catalyst affects the dynamics of a fixed bed reactor. Normally when the stability of a reactor is considered, problems arise due to the reaction rate accelerating, as for example the auto-ignition that occurs in exothermic reactions. Since catalyst deactivation results, in general, in a decrease of reaction rate, intuitively an increased stability would be expected when catalyst deactivation occurs. This, however, has not been found to be universally true, and in some cases peculiar trends arise due to catalyst deactivation which do not occur in deactivation free systems. Both experimental and simulation studies have confirmed that this unusual type of behaviour can occur.

7 Deactivation in Catalytic Reactors

The work of Ervin and Luss (1970) on the simulation of fixed bed adiabatic reactors subject to coking has already been referred to for the case when the catalyst pellets have a unique pseudo-steady state. This analysis also considered the possibility of non-unique pseudo-steady states for the catalyst pellet, and these were found to occur when the reactant partial pressure was increased from 0.07 to 0.16 (other data were the same as those listed in Table 7.1). When non-unique steady states occur the resultant profiles depend on the initial as well as the operating conditions. For the following examples the inlet temperature was fixed at 560°C.

Figures 7.22(a) and 7.22(b) portray the temperature and activity profiles respectively obtained by Ervin and Luss for parallel coking. A comparison of

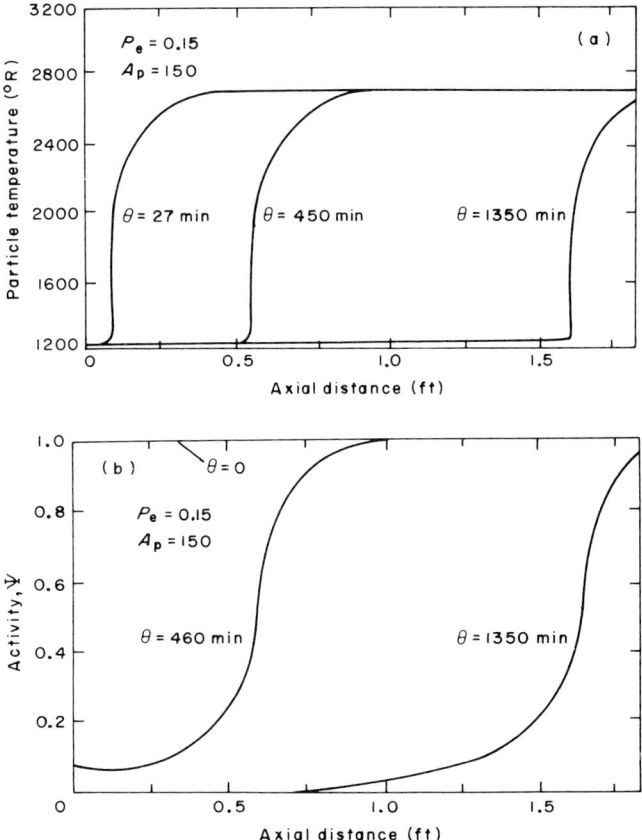

Fig. 7.22 (a) Transient temperature profiles for fouling by a parallel reaction for a case where some particles have non-unique pseudo-steady states. (b) Activity profiles for the case shown in (a). (Ervin and Luss, 1970)

Fig. 7.22(a) with Fig. 7.9 shows that the temperature profiles for the non-unique steady state do not spread out through the reactor with process time, but instead the reaction zone is confined to a few cells width for all times as it moves downstream. Deactivation occurs ahead of the reaction zone because in this region the concentration of reactant is high. The plug-like progression of the fouling is responsible for the downstream movement of the reaction zone; Fig. 7.23(b).

When fouling occurred by a series mechanism some interesting features were observed as shown in Fig. 7.23(a). The reaction zone did not move until 1500 min had elapsed, but at this point it moved very rapidly towards the bed exit and the temperature of the bed was in excess of the adiabatic

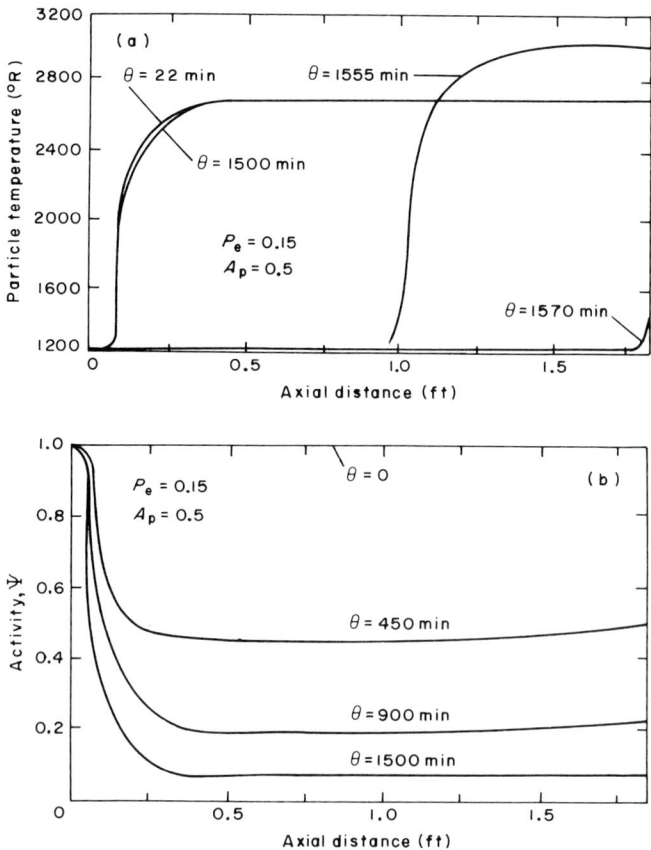

Fig. 7.23 (a) Transient temperature profiles for fouling by a consecutive reaction for a case where some particles have a non-unique pseudo-steady state. (b) Activity profiles for the case shown in (a). (Ervin and Luss, 1970)

temperature. At $\theta = 1570$ min the reaction zone was completely blown out of the reactor. This violent behaviour, which could seriously damage the catalyst pellets, is better understood by examination of the corresponding activity profile; Fig. 7.23(b). This shows that the region downstream from the reaction zone is fouled early in the process, and as the reaction zone moves downstream it encounters this region where coking has already occurred. The consequence is a rapid blowout of the reaction zone.

The influence of gas velocity variations was also investigated in this simulation. It was found that for parallel coking a step decrease in velocity caused a stabilization of the reaction zone. However, for series coking a velocity decrease caused the reaction zone to move upstream, while a velocity increase caused the reaction zone to move rapidly and completely out of the reactor. In this case the reactor was more sensitive to velocity fluctuations than it was when no deactivation occurred. These interesting results were not interpreted further by Ervin and Luss but they do demonstrate the unusual behaviour which can be obtained in certain instances when catalyst deactivation occurs.

Start-up is of particular importance in reactor operation, and some very interesting experimental results on the effect of fast deactivation on temperature profiles during start-up were obtained by Blaum (1974). Carbon monoxide oxidation over a nickel oxide catalyst was investigated in a pilot plant reactor with cooling at the wall. Excess air was used and the axial temperature measurements were recorded at different times during start-up of the reactor (Fig. 7.24). At zero time the temperature was constant over the whole reactor and equal to the inlet temperature which in these experiments was equal to the coolant temperature. A step increase in CO concentration at the inlet from 0 to 4.26 v/o led to a hot temperature front moving downstream. This front rapidly increased in temperature during the initial period, giving a maximum temperature "spike" at 166 min corresponding to a temperature rise of over 200°C. At 260 min the temperature profile had reached a steady state value with a temperature maximum of about 20°C above the inlet temperature. These large temperature rises could pose problems, and Blaum simulated his experimental results numerically using transient balances for heat and mass, and showed that the hot temperature front could be attributed to a fast deactivation process on the catalyst. For the particular case of CO oxidation used in this work, a catalytic reactivation process also occurs, and with appropriate choice of both deactivation and reactivation expressions, Blaum was able to obtain good agreement between experiment and simulation. However, it was necessary to assume that the time constant of the deactivation process and the time constant of the reactor, which is determined primarily by the heat capacity of the catalyst, were of the same order of magnitude.

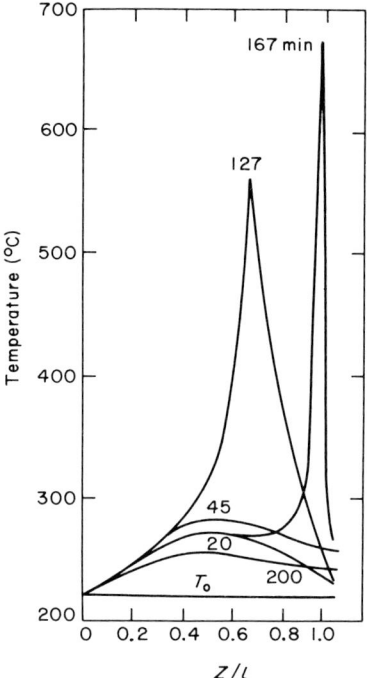

Fig. 7.24 Experimental axial temperature profiles from CO oxidation over a NiO catalyst. (Blaum, 1974)

The importance of thermal factors in the reactor has also been stressed by Billimoria and Butt (1981), who investigated experimentally the poisoning dynamics of a fixed bed. Thiophene was used to poison the nickel catalyst which was used to catalyse the hydrogenation of benzene. The bed thermal conductivity and heat capacity were varied by a factor of 2 using inert diluents in the catalyst bed. Such changes in the thermal properties of the bed were found to have a considerable influence on both the magnitude of the exotherms and the response of the system.

It would appear therefore that there are many unresolved and unusual features resulting from deactivation which occur during the dynamic stages of reactor operation. The existing literature is sparse on this topic and clearly much more work is required before definitive conclusions and predictions can be made in this area.

References

Anderson, R. B. and Whitehouse, A. M. (1961). *Ind. Eng. Chem.* **53**, 1011.
Billimoria, R. B. and Butt, J. B. (1981). *Chem. Eng. J.* **22**, 71.
Blaum, E. (1974). *Chem. Eng. Sci.* **29**, 2263.
Bohart, G. and Adams, E. (1920). *J. Amer. Chem. Soc.* **42**, 523.
Brito-Alayon, A., Hughes, R. and Kam, E. K. T. (1981). *Chem. Eng. Sci.* **36**, 445.
Brito-Alayon, A., Hughes, R. and Kam, E. K. T. (1982). *Chem. Eng. J.* **24**, 123.
Carberry, J. J. (1966). *Ind. Eng. Chem.* **58**, 40.
Deans, H. A. and Lapidus, L. (1960). *A.I.Ch.E.J.* **6**, 655.
Denny, P. J. and Twigg, M. W. (1980). In *Catalyst Deactivation* (ed. G. Froment and B. Delmon). Elsevier, Amsterdam.
Dumez, F. J. and Froment, G. F. (1976). *Ind. Eng. Chem. (Proc. Des. Devel.)* **15**, 291.
Eberly, P. E., Kimberlin, G. N., Miller, W. H. and Drushel, H. V. (1966). *Ind. Eng. Chem. (Proc. Des. Devel.)* **5**, 193.
Ervin, M. A. and Luss, D. (1970). *A.I.Ch.E.J.* **16**, 979.
Froment, G. F. and Bischoff, K. B. (1961). *Chem. Eng. Sci.* **16**, 189.
Froment, G. F. and Bischoff, K. B. (1962). *Chem. Eng. Sci.* **17**, 105.
Haynes, H. W., Jr. (1970). *Chem. Eng. Sci.* **25**, 1615.
Kam, E. K. T. and Hughes, R. (1979a). *A.I.Ch.E.J.* **25**, 359.
Kam, E. K. T. and Hughes, R. (1979b). *Chem. Eng. J.* **18**, 93.
Kam, E. K. T., Ramachandran, P. A. and Hughes, R. (1975). *J. Catal.* **38**, 283.
Kam, E. K. T., Ramachandran, P. A. and Hughes, R. (1977a). *Chem. Eng. Sci.* **32**, 1307, 1317.
Kam, E. K. T., Ramachandran, P. A. and Hughes, R. (1977b). *J. Catal.* **48**, 177.
Koch, W. H., Greenberg, D. B., Blanchard, J. A. and Ramaswami, K. (1980). *Chem. Eng. Commun.* **4**, 759.
Lee, J. W. and Butt, J. B. (1973). *Chem. Eng. J.* **6**, 111.
Masamune, S. and Smith, J. M. (1966). *A.I.Ch.E.J.* **12**, 384.
McGreavy, C. and Cresswell, D. L. (1969). *Can. J. Chem. Eng.* **47**, 583.
Moseley, F., Stephens, R. W., Stewart, K. D. and Wood, J. (1972). *J. Catal.* **24**, 18.
Olson, J. H. (1968). *Ind. Eng. Chem. (Fund.)* **7**, 185.
Ozawa, Y. and Bischoff, K. B. (1968). *Ind. Eng. Chem. (Proc. Des. Devel.)* **7**, 65, 72.
Valdman, B. (1976). Ph.D. Thesis, Salford University.
Van Zoonen, D. D. (1965). *Proc. 3rd Int. Congr. Catalysis*, p. 1316. North-Holland Publishing Co.
Vermeulen, T. (1958). *Adv. Chem. Eng.* **2**, 148.
Vermeulen, T. (1963). *Chem. Eng. Handbook* (ed. J. H. Perry), 4th Edn. McGraw-Hill, New York.
Weng, H. S., Eigenberger, G. and Butt, J. B. (1976). *Chem. Eng. Sci.* **31**, 1341.
Wheeler, A. (1955). In *Catalysis*, Vol. II (ed. P. H. Emmett).
Wheeler, A. and Robell, A. J. (1969). *J. Catal.* **13**, 299.

8
Optimization of Deactivating Reactor Systems

8.1 Comparison of various reactor types under deactivating conditions

In its widest sense optimization includes not only the best operating procedure for a given reactor type but also the choice of reactor system. As noted in the preceding chapter, the choice of reactor can be important when it is desired to minimize problems arising from catalyst deactivation.

In many instances the nature of the deactivation process determines which type of reactor should be employed. An obvious example is the use of fluid bed or transfer line reactors when rapid coking occurs. This enables a solids flow system to be used for continuous regeneration of spent catalyst; if this facility was not available the system would soon become inoperable due to the rapid accumulation of coke.

In other cases, however, the rate of deactivation may be much slower and there may also be other reasons why a particular type of reactor may be preferred or rejected. Thus, attrition of catalysts in fluid bed reactors may preclude the use of these for a particular catalytic process.

Although Levenspiel (1972) devised methods by which the rate constants for the deactivation process may be obtained in various types of reactors, he did not compare the various reactors as such, in terms of performance under these conditions. Moreover, the analysis was limited to conditions where the deactivation process was either independent of gas phase concentration or operated with a constant concentration of the deactivating precursor. The former corresponds to thermal sintering alone and is therefore of limited application particularly since these processes are usually very slow and it is difficult to obtain accurate laboratory data for long sintering times. Batch, stirred tank, and plug flow reactors were considered by Levenspiel in this work.

For gas or vapour phase processing the choice of reactor is generally limited to a fixed bed, or a moving bed, or a fluid bed system. The simple correlation of Voorhies (1945) was employed by Andrews (1959) to compare

8 Optimization of Deactivating Reactor Systems

the performances of various reactor types. Subsequently, although a great deal of work was reported on catalyst deactivation, no assessment was made of different reactor types subject to catalyst fouling until Weekman (1968) proposed a mathematical model for catalytic cracking under conditions of catalyst deactivation in fixed, moving, and fluid bed reactors. Weekman noted that these reactions have generally been observed to obey second order kinetics while the rate of catalyst deactivation has been correlated with a simple first order decrease with time of utilization.

The work of Weekman was extended by Sadana and Doraiswamy (1971) who considered a main reaction of generalized order and also the effect of a deactivation process which could take various forms. Weekman considered an exponential activity function only, similar to that adopted by Froment and Bischoff (1962). The treatment which follows is based upon that adopted by Sadana and Doraiswamy.

In a fixed bed reactor operating under isothermal conditions, and in which a reactant A is fed and reacts under mass transfer free conditions, the continuity equation assuming plug flow is:

$$\varepsilon \frac{\rho_g}{M}\left(\frac{\partial y}{\partial t}\right) + \frac{G}{M}\left(\frac{\partial y}{\partial z}\right) = r_A(y, t)\varepsilon \qquad (8.1)$$

where G is the mass flow rate of reactant of molecular weight M, density ρ_g, and mole fraction y. It should be noted that the reaction rate r_A is a function of both the mole fraction of A and the reaction time t.

For a fixed bed reactor the time on stream, t_d, is used to normalize the time at any point during the reaction or decay cycle. Then the governing equation in normalized coordinates is:

$$A'\left(\frac{\partial y}{\partial t}\right) + \left(\frac{\partial y}{\partial \xi}\right) = -Br_A(y, \bar{t}) \qquad (8.2)$$

where

$$\bar{t} = \frac{t}{t_d}; \quad \xi = \frac{z}{L}; \quad B = \frac{\varepsilon \rho_g}{\rho_F S_F} \qquad (8.3)$$

and

$$A' = \frac{\varepsilon \rho_g}{\rho_F S_F t_d}$$

where ρ_F is the density of any feed and S_F is the space velocity of the feed.

The term G has now been replaced by $\rho_F S_F L$, and A' represents the ratio of the feed transit time through the reactor to the catalyst decay time. This is generally of negligible magnitude and, assuming $A' = 0$, equation (8.2) becomes:

$$\left(\frac{dy}{d\xi}\right) = -Br_A(y, \bar{t}) \tag{8.4}$$

The rate of disappearance of A for kinetics of any order, m, is:

$$r_A = k_0 \, e^{-\lambda \bar{t}} y^m \tag{8.5}$$

where k_0 is the rate constant at zero time and:

$$\lambda = \alpha(t_d)^d$$

$$= \alpha t_d \quad \text{(for first order decay)} \tag{8.6}$$

represents the activity function.

Substituting equation (8.5) into (8.4) we get:

$$\frac{dy}{d\xi} = -B' \, e^{-\lambda \bar{t}} y^m \tag{8.7}$$

where

$$B' = Bk_0 \tag{8.8}$$

and can be regarded as a reaction group.

The solution of equation (8.7) is

$$y = \left[\frac{1}{(m-1)B'\xi \, e^{-\lambda \bar{t}} + 1}\right]^{1/(m-1)} \quad (m \neq 1) \tag{8.9}$$

$$y = \exp(-B'\xi \, e^{-\lambda \bar{t}}) \quad (m = 1) \tag{8.10}$$

The average conversion obtained at the reactor outlet ($\xi = 1$) is:

$$X_A = 1 - \int_0^1 y \, d\bar{t} \tag{8.11}$$

Introducing equations (8.9) and (8.10) into (8.11) we get:

mth order:

$$X = 1 - \int_0^1 \left[\frac{1}{(m-1)(B'\xi \, e^{-\lambda \bar{t}} + 1)}\right]^{1/(m-1)} d\bar{t} \tag{8.12}$$

1st order:

$$X = 1 + \frac{1}{\lambda}[E_i^*(B') - E_i^*(B') \, e^{-\lambda}] \tag{8.13}$$

where

$$E_i^*(h) = -E_i(-h) = \int_h^\infty \frac{e^{-t}}{t} \, dt$$

is the exponential integral.

8 Optimization of Deactivating Reactor Systems

Similarly, equations for conversion can also be derived for the linear deactivation expression represented by:

$$k(\bar{t}) = k_0 - \lambda \bar{t} \tag{8.14}$$

and another alternative deactivation expression:

$$k(\bar{t}) = k_0 - \lambda \bar{t}^d \tag{8.15}$$

For a moving bed reactor at steady state, the residence time of the decaying catalyst corresponds to the total decay time, t_d, of the fixed bed reactor. Therefore:

$$\frac{dy}{d\xi} = -B' e^{-\lambda \xi} y^m \tag{8.16}$$

and the solution of equation (8.16) is:

$$y = \left[\frac{\lambda}{(m-1)B'(1-e^{-\lambda \xi}) + \lambda}\right]^{1/(m-1)} \quad (m \neq 1) \tag{8.17}$$

$$y = \exp\left[\frac{B'}{\lambda}(e^{-\lambda \xi} - 1)\right] \quad (m = 1) \tag{8.18}$$

For steady state operation of the moving bed reactor:

$$X = 1 - y \tag{8.19}$$

and substituting equations (8.17) and (8.18) into (8.19) we obtain at the reactor exit ($\xi = 1$):

mth order:

$$X = 1 - \left[\frac{\lambda}{(m-1)B'(1-e^{-\lambda}) + \lambda}\right]^{1/(m-1)} \tag{8.20}$$

1st order:

$$X = 1 - \exp\left[\frac{B'}{\lambda}(e^{-\lambda} - 1)\right] \tag{8.21}$$

Use of other deactivation expressions will give other solutions.

In the case of a fluid bed reactor the gas is assumed to be in plug flow and the solid phase is assumed to be completely mixed. For a perfectly mixed catalyst the age distribution is given by $\exp(-\bar{t}_d)$. Weekman (1968) has given the value of the rate constant for this case as:

$$[k(\bar{t})]_{Av} = \frac{k_0}{1 + \lambda} \tag{8.22}$$

Combining this with equations (8.4) and (8.8) gives:

$$\frac{dy}{d\xi} = -\frac{B'}{1+\lambda} y^m \tag{8.23}$$

The solution of this is:

*m*th order:

$$y = \left[\frac{\lambda + 1}{(m-1)B'\xi + (\lambda + 1)}\right]^{1/(m-1)} \tag{8.24}$$

1st order:

$$y = \exp\left(-\frac{B'\xi}{\lambda + 1}\right) \tag{8.25}$$

and the corresponding conversions at the reactor exit are:

*m*th order:

$$X = 1 - \left[\frac{(\lambda + 1)}{(m-1)B' + (\lambda + 1)}\right]^{1/(m-1)} \tag{8.26}$$

1st order:

$$X = 1 - \exp\left(-\frac{B'}{\lambda + 1}\right) \tag{8.27}$$

As for the other two reactor types, alternative deactivation functions may be used. Table 8.1 summarizes the various solutions for the three different reactor types using different deactivation expressions. Equations are given both for the reactant and for the conversion.

As noted above, Weekman (1968) has compared fixed, fluid, and moving bed reactors for the special case of cracking reactions and using an exponential activity function. For cracking reactions the order of reaction is generally accepted as 2, and using this Weekman produced comparative plots of the three reactor types. In Fig. 8.1 the ratio of fixed bed to moving bed conversion is plotted against the parameter B' for different extents of catalyst decay as measured by the deactivation parameter λ. The parameter B' is essentially a measure of the extent of reaction since it represents the ratio of specific reaction velocity to the residence time of the gas or vapour stream. Figure 8.1 indicates that, under any conditions of catalyst decay, moving beds give a higher conversion than fixed beds, which is to be expected in view of the different catalyst residence times in the two types of reactor. However, it should be borne in mind that this conclusion assumes that the moving bed catalyst is always completely regenerated before re-entering the reactor; this may not always be the case in practice. Moving bed and fluid bed conversions are compared in Fig. 8.2(a) and fixed beds to fluid beds in Fig.

8 Optimization of Deactivating Reactor Systems

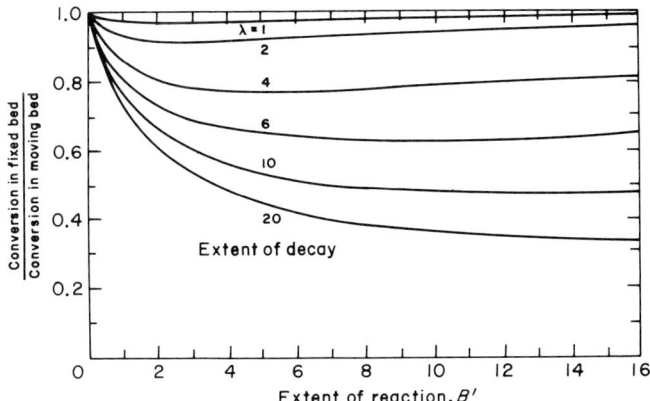

Fig. 8.1 Ratio of fixed bed to moving bed conversion. (Weekman, 1968)

8.2(b). In these figures the ratio of conversion is now plotted against λ, with B' as parameter on the curves. Figure 8.2(a) demonstrates that the moving bed has a higher conversion than a fluid bed, and this difference between the two reactor types is greatest for slow rates of deactivation (λ small). However, this difference becomes very small at large extents of reaction or when the deactivation parameter, λ, becomes larger. The comparison of fixed and fluid bed reactors in Fig. 8.2(b) is somewhat more complicated. When the extent of reaction is small and λ is also small (slow deactivation) a fixed bed is preferable but as both of these parameters increase the fluid bed is preferred. These results are in accord with practice where fixed beds are used for slow deactivation processes while fluid beds are employed for reactions where the catalyst is rapidly coked.

The above analysis of Weekman has been extended by Sadana and Doraiswamy (1971) to consider reactions of any order and the effects of axial dispersion. The effect of reaction order on the three reactor types is illustrated in Fig. 8.3. When the deactivation parameter λ is zero, the equations for all three reactor types reduce to the same value for conversion, regardless of the deactivation expression employed. This is shown by the single line at $\lambda = 0$ in Fig. 8.3. When finite values of the decay group λ are employed, however, different curves are obtained for fixed, moving, and fluid bed reactors. Figure 8.3 presents curves for $\lambda = 6$ and for $\lambda = 10$. The effect of deactivation is clearly seen to be far less severe for reactions of higher order in a moving or fluid bed reactor than in a fixed bed reactor. For the latter the effect of reaction order is virtually negligible. It may be concluded therefore that moving bed and fluid bed reactors, which are generally superior to the fixed bed reactor when catalyst deactivation by fouling occurs, should be

Table 8.1 Expressions for mole fraction and conversion A for various decay forms.

Decay form	Expressions for y_A mth order (m ≠ 1)	1st order
$k = k_o e^{-\lambda \bar{t}}$		
Fixed	$\left[\dfrac{1}{(m-1)B' e^{-\lambda \bar{t}} + 1}\right]^{1/(m-1)}$	$\exp[-B' e^{-\lambda \bar{t}}]$
Moving	$\left[\dfrac{\lambda}{(m-1)B'(1 - e^{-\lambda}) + \lambda}\right]^{1/(m-1)}$	$\exp\left[\dfrac{B'}{\lambda}(e^{-\lambda} - 1)\right]$
Fluid	$\left[\dfrac{(\lambda + 1)}{(m-1)B' + (\lambda + 1)}\right]^{1/(m-1)}$	$\exp\left[-\left(\dfrac{B'}{\lambda + 1}\right)\right]$
$k = k_o - \lambda \bar{t}$		
Fixed	$\left[\dfrac{k_o}{B'(m-1)(k_o - \lambda \bar{t}) + k_o}\right]^{1/(m-1)}$	$\exp\left[-\left(B' - \dfrac{B' \lambda \bar{t}}{k_o}\right)\right]$
Moving	$\left[\dfrac{2k_o}{B'(m-1)(2k_o - \lambda) + 2k_o}\right]^{1/(m-1)}$	$\exp\left[B'\left(\dfrac{\lambda}{2k_o} - 1\right)\right]$
Fluid	$\left[\dfrac{k_o}{B'(k_o - \lambda)(m-1) + k_o}\right]^{1/(m-1)}$	$\exp\left[\dfrac{B'}{k_o}(k_o - \lambda)\right]$
$k = k_o - \bar{t}^d$		
Fixed	$\left[\dfrac{k_o}{B'k_o - B' \lambda \bar{t}^d (m-1) + k_o}\right]^{1/(m-1)}$	$\exp\left[\left(\dfrac{B'}{k_o}\lambda \bar{t}^d - B'\right)\right]$
Moving	$\left[\dfrac{(d+1)k_o}{(m-1)B'[(d+1)k_o - \lambda] + (d+1)k_o}\right]^{1/(m-1)}$	$\exp\left[B'\left(\dfrac{\lambda}{k_o(d+1)} - 1\right)\right]$
Fluid	$\left[\dfrac{k_o}{k_o + (m-1)B'(k_o - d\lambda)}\right]^{1/(m-1)}$	$\exp\left[B'\left(\dfrac{d\lambda}{k_o} - 1\right)\right]$

8 Optimization of Deactivating Reactor Systems

Table 8.1 *cont.*

Expressions for conversion	
mth order ($m \neq 1$)	1st order

$$1 - \int_0^1 \left[\frac{1}{(m-1)B'e^{-\lambda \bar{t}_p} + 1} \right]^{1/(m-1)} d\bar{t} \qquad\qquad 1 + \frac{1}{\lambda} E_i^*(B') - E_i^*(B'e^{-\lambda})$$

$$1 - \left[\frac{\lambda}{(m-1)B'(1-e^{-\lambda}) + \lambda} \right]^{1/(m-1)} \qquad\qquad 1 - \exp\left[\frac{B'}{\lambda}(e^{-\lambda} - 1) \right]$$

$$1 - \left[\frac{(\lambda + 1)}{(m-1)B' + (\lambda + 1)} \right]^{1/(m-1)} \qquad\qquad 1 - \exp\left[-\left(\frac{B'}{\lambda + 1}\right) \right]$$

$$1 - \int_0^1 \left[\frac{k_o}{B'(m-1)(k_o - \bar{t} + k_o)} \right]^{1/(m-1)} d\bar{t} \qquad\qquad 1 - \int_0^1 \exp\left[-\left(B' - \frac{B'\lambda \bar{t}}{k_o}\right) \right] d\bar{t}$$

$$1 - \left[\frac{2k_o}{(m-1)B'(2k_o - \lambda) + 2k_o} \right]^{1/(m-1)} \qquad\qquad 1 - \exp\left[B'\left(\frac{\lambda}{2k_o} - 1\right) \right]$$

$$1 - \left[\frac{k_o}{B'(k_o - \lambda)(m-1) + k_o} \right]^{1/(m-1)} \qquad\qquad 1 - \exp\left[-\frac{B'}{k_o}(k_o - \lambda) \right]$$

$$1 - \int_0^1 \left[\frac{k_o}{(B'k_o - B'\lambda \bar{t}^d)(m-1) + k_o} \right]^{1/(m-1)} d\bar{t} \qquad\qquad 1 - \int_0^1 \exp\left[\frac{B'}{k_o} \lambda \bar{t}^d - B' \right] d\bar{t}$$

$$1 - \left[\frac{(d+1)k_o}{(m-1)B'[(d+1)k_o - \lambda] + (d+1)k_o} \right]^{1/(m-1)} \qquad\qquad 1 - \exp\left[B'\left(\frac{\lambda}{k_o(d+1)} - 1\right) \right]$$

$$1 - \left[\frac{k_o}{k_o + (m-1)B'(k_o - d\lambda)} \right]^{1/(m-1)} \qquad\qquad 1 - \exp\left[B'\left(\frac{d\lambda}{k_o} - 1\right) \right]$$

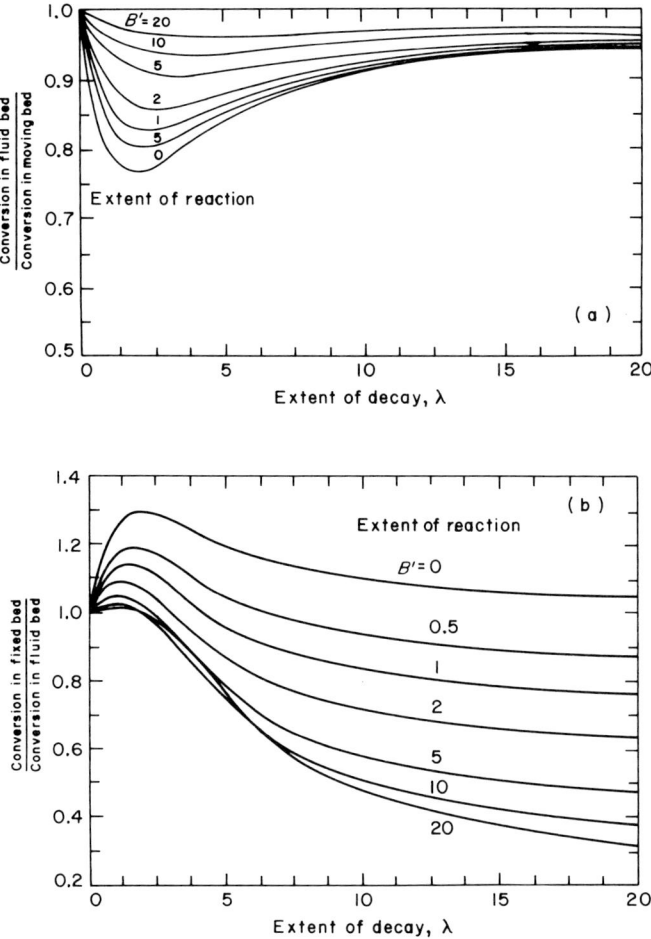

Fig. 8.2 Ratio of (a) fluid bed to moving bed conversion and (b) fixed bed to fluid bed conversion. (Weekman, 1968)

preferred when reactions of higher order are being processed under conditions of severe fouling.

The effect of axial dispersion was also considered by Sadana and Doraiswamy. It was observed that, at relatively low values of the reaction group B', the increased conversion which occurs under plug flow conditions almost disappears as the decay parameter increases.

It must be stressed that the simplifying assumptions on which the various models are based must be borne in mind when the above comparisons are

8 Optimization of Deactivating Reactor Systems

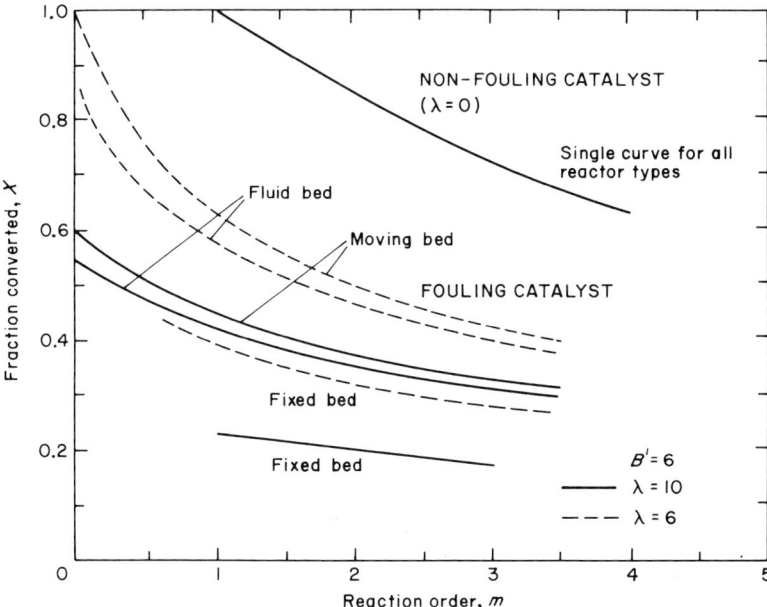

Fig. 8.3 Fractional conversion as a function of the reaction order for the three reactor types at relatively low and high values of the fouling factor. (Sadana and Doraiswamy, 1971)

made. Incorporation of non-isothermal conditions and/or diffusional and mass transfer resistances could modify the conclusions drawn above.

8.2 Optimization of deactivating reactors

There has been a great deal of interest in this topic over the past 10–15 years, which is reflected in the large number of theoretical papers published on this topic. As pointed out by Park and Levenspiel (1976), operation of a reactor with decaying catalyst involves two problems: (1) how best to operate the reactor during a run (the operational problem); (2) when to stop a run and regenerate or replace the catalyst (the regeneration problem).

The regeneration problem has attracted far less attention than the operational problem. The regeneration phase of the reaction/regeneration cycle is strongly dependent on the temperature progression during the operational phase. This essential dependence has been demonstrated in a numerical example by Miertschin and Jackson (1970).

A number of variables enter into the general problem of optimizing reactor performance during deactivation. These include catalyst composition and distribution, reactor type and size, temperature of operation and conversion, and selectivity constraints. The question of reactor type has already been discussed in Section 8.1, so further discussion follows on the effect of the other variables.

8.2.1 Optimal temperature policies

As has been recognized commercially for many years, the most important independent variable is temperature, and thus the essential problem is to find this optimum temperature policy. The way in which this optimum temperature policy is formulated is important; usually the aim is to keep the exit conversion constant so that the overall reactor temperature is increased to compensate for the loss in activity of the catalyst by increasing the rate constant. This has been practised industrially by analysing the exit stream and/or measuring the temperature profile in the reactor. The loss in conversion is offset by increasing the temperature as illustrated in Fig. 8.4 where bed temperature profiles for a low temperature shift catalyst are illustrated. When the catalyst charge is fresh the reaction starts at the bed inlet and the temperature profile through the catalyst bed has the form shown by curve A of Fig. 8.4. With a new catalyst the maximum rate of temperature rise should be at the bed inlet and the rate of rise at the exit should be very slow because the gas composition is close to equilibrium. Curve B shows the temperature profile at the middle of the catalyst life. There is no temperature rise at the bed inlet and to maintain the activity the temperature has been raised slightly. Curve C shows a typical temperature profile when the catalyst is almost spent. The inlet temperature has now been raised appreciably in order to obtain the best possible performance, to compensate for the absence of reaction in a large portion of the bed. When reaction starts it is still rapid but the exit gas is further from equilibrium than in curves A and B and the catalyst requires renewing.

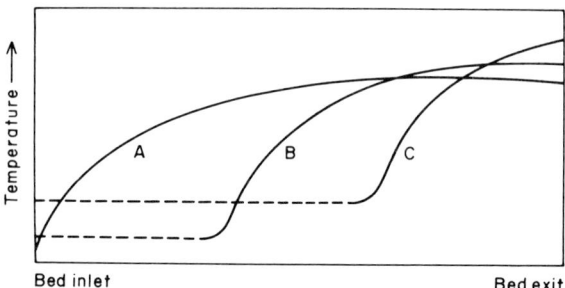

Fig. 8.4 Temperature profile through catalyst bed: (A) new; (B) mid-life; (C) end life. (Young and Clark, 1973)

The objective function usually specified in theoretical studies has in general had the aim of formulating precisely what may be done by experience in practical situations. Thus, the aim has been to maximize the final conversion for a fixed time of reaction and specified final catalyst activity by varying the temperature with time on stream, subject to upper and lower bounds on the allowable temperature. This has been done by Szépe and Levenspiel (1968) for batch reactors and by Chou et al. (1967) for tubular reactors. In both cases the conversion independent form of deactivation kinetics was employed. This means that only independent poisoning or fouling by an impurity in the feed stream or sintering are possible deactivation processes.

In their work Szépe and Levenspiel determined some important consequences of such an optimum temperature policy. The first, and perhaps the most obvious, is that the effective rate constant remains unchanged during the entire process. This is applicable to a reaction of any order and any deactivation order (a deactivation expression of the form $\psi(S) = S^m$ was used, where m is the deactivation order). It was also found that the optimal temperature policy is only applicable if the activation energy of the deactivation process, E_d, is less than that of the main reaction, E. If in any case $E_d < E$, obviously the temperature should be kept as high as is practically feasible during the entire process, and the problem then becomes trivial. For equal values of the activation energies all policies will be the same.

For $E_d < E$ the optimal condition was expressed by Szépe and Levenspiel in the form of the following equations for temperature and activity:

$$\frac{d(1/T)}{dt} = -k_d^o \frac{R}{E} \exp\left(-\frac{E_d}{RT}\right) S^{m-1} \tag{8.28}$$

$$\frac{dS}{dt} = -k_d^o \exp\left(-\frac{E_d}{RT}\right) S^m \tag{8.29}$$

which together with the conditions:

$$\left.\begin{array}{l} S(0) = S_o \\ S(t_e) = S_e \end{array}\right\} \tag{8.30}$$

constitute the temperature policy.

For first order deactivation ($m = 1$) equations (8.28) and (8.29) can be solved separately and the final expression for the reactor temperature as a function of time is:

$$T = \frac{E_d}{R} \ln \left\{ \frac{E_d}{E} k_d^o t_e \left[1 - \left(\frac{S_e}{S_o}\right)^{E_d/E} \right]^{-1} - \tau \right\} \tag{8.31}$$

where $\tau = t/t_e$ and t_e is the duration of the cycle.

This equation is of the form:

$$T = \frac{1}{A + B\ln(C - \tau)} \qquad (8.32)$$

and indicates that the optimal temperature policy is a monotonically increasing function of time with the slope increasing with time. Thus the optimal temperature policy has the form shown in Fig. 8.5(a).

The above assumes no upper temperature constraint. Since temperature can only be positive, equation (8.31) has a physical meaning only if the argument of the logarithmic term is greater than unity. Therefore this implies that:

$$\frac{S_e}{S_o} \geqslant \left[\frac{1}{1 + (E_d/E)k_d^o t_e}\right]^{E/E_d} \qquad (8.33)$$

Furthermore, if a maximum allowable temperature, T_{max}, is specified, the constraint of equation (8.33) corresponds to:

$$\frac{S_e}{S_o} \geqslant \left[\frac{1}{1 + (E_d/E)k_d^o t_e \exp(-E_d/RT_{max})}\right]^{E/E_d} \qquad (8.34)$$

If the parameters of the problem do not satisfy the inequality of (8.34), the optimal policy will be one of two limiting cases. If

$$\exp\left[-k_d^o t_e \exp\left(\frac{E_d}{RT_{max}}\right)\right] \leqslant \frac{S_e}{S_o} \leqslant \left[\frac{1}{1 + (E_d/E)k_d^o t_e \exp(-E_d/RT_{max})}\right]^{E/E_d} \qquad (8.35)$$

the optimal temperature policy will trace a rising temperature curve followed by a constant temperature plateau as shown in Fig. 8.5(b).

Secondly, if

$$\frac{S_e^*}{S_o} \leqslant \exp[-k_d^o t_e \exp(-E_d/RT_{max})] \qquad (8.36)$$

where S_e^* denotes the specified final activity, the optimal policy will be to operate throughout at the maximum allowable temperature as in Fig. 8.5(c). This is similar to the situation when $E_d < E$.

More recently, Sadana and Levenspiel (1978) have extended the above analysis to show by means of a worked example that the optimum temperature policy (for $E_d > E$) will be as in Fig. 8.5(a) even if there is a temperature constraint. This analysis was for a packed bed reactor but the same problem formulation was used as for the batch reactor. Sadana and Levenspiel argued that, if a temperature constraint is present, the optimum will follow a constant conversion, rising temperature path as in Fig. 8.5(a) and ending the run as the temperature just reaches the maximum allowable

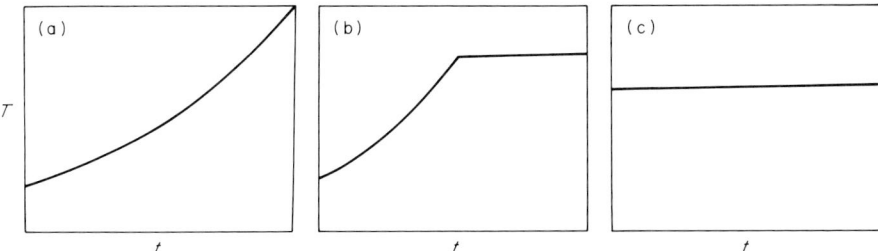

Fig. 8.5 Optimal temperature policy in a batch reactor with an upper temperature limit.

temperature. Operation as in Fig. 8.5(b) will not be optimal because of the declining conversion with time once this maximum temperature has been reached. For best operation it would be better to start at a lower temperature so as to arrive at the maximum allowable temperature, T^*, as the run is terminated.

Although the above analysis predicts the optimal temperature policy for concentration independent deactivation processes, Lee and Crowe (1970) have warned that it cannot be applied to reactions where deactivation is concentration dependent, such as coking processes. They showed that, if this were attempted, the inclusion of the concentration dependent activity would reduce the rate of rise of temperature with time in comparison with the concentration independent case. In the limit this could even lead to a decreasing temperature policy with time.

A similar approach to that of Szépe and Levenspiel (1968) was employed by Chou *et al.* (1967) but in this case tubular reactors were analysed. If plug flow is assumed the reactor mass balance is:

$$\frac{\partial x}{\partial t} + u \frac{\partial x}{\partial z} = r_A \tag{8.37}$$

in which x is the extent of reaction and u the superficial gas velocity. The rate of the main reaction is a function of temperature T, x, and activity S. Therefore:

$$\frac{\partial a}{\partial t} = r_{Ad}(x, S, T) \tag{8.38}$$

The objective function has the aim of maximizing the total yield, Y, of the reactor:

$$Y = \int_0^{t_e} x(z, t) \, dt \tag{8.39}$$

for a specified time by varying the temperature profile.

In the simplest case, isothermality within the reactor is assumed and the rate of catalyst deactivation depends on temperature but is again assumed independent of concentration (or x). Furthermore, pseudo-steady state conditions are assumed so that the first term in equation (8.37) may be neglected. Maximizing equation (8.37) gives the condition:

$$\left(\frac{z}{u}\right) S k^\circ \exp(-E/RT) = \text{const.} = \sigma = \ln\left(\frac{1}{1-x_z}\right) \quad (8.40)$$

when the deactivation kinetics are expressed as:

$$\frac{dS}{dt} = -k_d^\circ S \exp(-E_d/RT) \quad (8.41)$$

Equation (8.40) can be written as:

$$-\left(\frac{E_d}{E}\right) S^{(E_d/E-1)} \frac{dS}{dt} = \frac{k_d^\circ E_d}{E}\left(\frac{\sigma}{k_d^\circ \frac{z}{u}}\right)^{E_d/E} \quad (8.42)$$

Solving for $S(t)$:

$$S(t) = (1 - \beta t)^{E/d} \quad (8.43)$$

where β is the right-hand side of equation (8.42). Solving equation (8.41) for $S(t)$ and writing the temperature dependence directly into equation (8.43), the optimum temperature policy is:

$$T(t) = \left(\frac{E_d}{R}\right) \ln[k_d^\circ(1 - \beta t)/\beta E_d/E] \quad (8.44)$$

The initial temperature of operation follows from this by putting $t = 0$ and is:

$$T(0) = \left(\frac{E_d}{R}\right) \ln[k_d^\circ(E_d/E)] \quad (8.45)$$

Combination of equations (8.44) and (8.45) gives:

$$\frac{E_d}{RT(0)} - \frac{E_d}{RT(t)} = \ln\left(\frac{1}{1-\beta t}\right) \quad (8.46)$$

and this enables the true temperature policy to be determined up to some specified limit on reactor temperature:

$$t_e = \frac{1}{\beta}\left\{1 - \exp\left[-\frac{E}{R}\left(\frac{1}{T(0)} - \frac{1}{T_{\max}}\right)\right]\right\} \quad (8.47)$$

The total yield will then be

$$Y = x_z t_e \tag{8.48}$$

and the maximum extent of reaction obtainable with fresh catalyst at $T_{\max}$ is defined as:

$$x_z^* = 1 - \exp\left(-k^* \frac{z}{u}\right) \tag{8.49}$$

The dimensionless yield is defined as $\eta = k_d^* Y$ and can be written as:

$$\eta = \frac{E}{E_d} x_z \left\{ \left[\frac{\ln\left(\frac{1}{1-x_z^*}\right)}{\ln\left(\frac{1}{1-x_z}\right)} \right]^{E_d/E} - 1 \right\} \tag{8.50}$$

Some results obtained by Chou et al. are shown in Fig. 8.6. Here the activation energy ratio E/E_d is plotted against $(1 - x_z^*)$ which is the optimal steady state conversion, with the extent of conversion x as parameter. To maximize the yield Y, different procedures apply according to the relative magnitude of the activation energies for reaction and deactivation. If $E < E_d$ the yield will increase monotonically as x decreases [from equation (8.44)]. Thus, x should be made as small as possible to give the greatest yield. If $E > E_d$ there is a value of x at which the dimensionless yield η assumes its maximum possible value. Figure 8.6 shows how the constant conversion should be chosen given E/E_d and x_z^* (i.e. $T_{\max}$). It can also be seen from Fig. 8.6 that high levels of conversion are only optimal when the ratio of activation energies is large and the maximum conversion is close to 100%. The optimal reactor size may also be estimated similarly. A balance has to be struck in this case between increasing cost with increasing size of reactor against increased productivity.

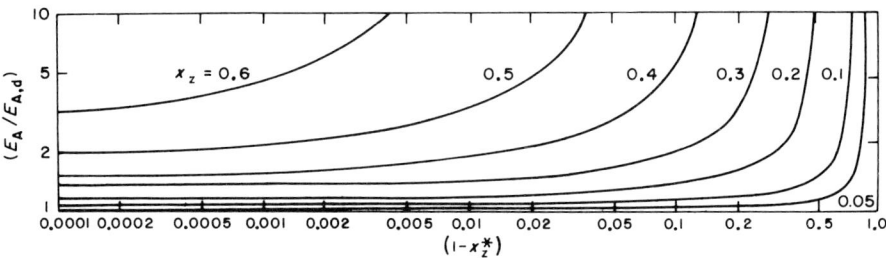

Fig. 8.6 Optimal steady conversion for given activation energy ratio and maximum extent. (Chou et al., 1967)

Other analyses of temperature optimization have been made by numerical methods using Pontryagin's maximum principle by Ogunye and Ray (1968). The problem considered was the series reaction A → B → C which occurs in an isothermal reactor with a monofunctional catalyst. The temperature is plotted against operating time to maximize the production of the desired intermediate B. This again is subject to a maximum temperature constraint.

Considerable effort is currently being expended in this area of optimal temperature policies for deactivating reactor systems, with particular emphasis on fixed bed reactors. The complexity of the problems which relate to actual industrial practice is undoubtedly very great and this presents a very challenging task.

8.2.2 Other optimal policies

Although most attention has been directed to optimal temperature policies, presumably because these are ones which can be operated most readily in practice, other variables have been considered in this context of maximizing conversion.

One interesting study is that of DeLancey (1973) who considered the optimum structure of the catalyst pellet and optimum distribution of the active ingredient. For any catalyst the maximum reaction rate is achieved by uniformly impregnating with the active ingredient at the maximum level, provided of course that the pore-surface area characteristics of the pellet are not adversely affected. However, under deactivating conditions when catalyst replacement or regeneration must be considered, a competition between catalyst economics and activity arises. DeLancey studied the possibility of decreasing the operating expenses associated with the problem of catalyst deactivation by optimizing the distribution of active material in the catalyst pellet for fixed operating conditions which, in this context, means temperature and time on stream. The analysis assumed a homogeneous type poisoning by an independent poison with separable kinetics for the main and poisoning reactions. The effective rate constant was written in the form:

$$k = k_o \exp(-t/\tau_p) \tag{8.51}$$

where k_o is the unpoisoned value of the rate constant and t and τ_p are the operating time and poisoning time constant respectively. The latter was approximated by the Langmuir–Hinshelwood expression for the surface concentrations:

$$\tau_p \approx 1 + (1 + K_A C_{Ao} + K_B C_{Bo})/K_P C_{Po} \tag{8.52}$$

where K_i is the absorption equilibrium constant for species i, and B and P denote the product and poison respectively.

The rate constant k_o is usually taken to be representative of conditions when the maximum number of sites in the catalyst are activated. For the calculations made by DeLancey k_o refers to catalyst depositions below this maximum and is therefore represented as:

$$k_o = k^o b \tag{8.53}$$

where b is the mass of catalyst supported on unit area of support. The proportionality factor k_o is taken as constant, i.e. it is assumed that activity is linear in b up to the maximum value b_{max}.

Assuming the usual pseudo-steady state conditions hold, the mass balance may be solved for the function u ($u = b/b_{max}$), and with appropriate cost factors allowing for net return on products made and catalyst replacement costs, an objective function may be defined as the net return to be expected for a pellet operating at an average deactivation level for a time θ.

DeLancey found that there was an optimal impregnation policy for the active ingredient in catalyst pellets. This optimal policy is due to the recurring catalyst costs and to concentration gradients within the pellet which decrease the extent to which active material is used. These concentration gradients are caused by the diffusional resistances as isothermal conditions are assumed. However, non-isothermal effects were also included in a development of the isothermal case.

The analysis was also extended to a fixed bed reactor undergoing deactivation using the same optimal policy. Uniform poisoning was assumed under isothermal conditions in the absence of axial dispersion. A uniform activation policy that was optimal for the packed bed was calculated. This policy depends only on the cost factor for the catalyst, the Thiele modulus of the pellet, the dimensionless time, and the ratio θ/τ_p. Although this method of optimal operation may be difficult to achieve in practical cases, it is interesting to observe that variations in the distribution of the active material can lead to better operating practice.

An important factor in the design of a fixed bed reactor operating under conditions where catalyst deactivation occurs is the optimization of the time on stream. The lower the value of this time the greater is the frequency of regeneration and the greater the operational costs. In consecutive reactions of the type

$$A \xrightarrow{k_1} R \xrightarrow{k_2} S$$

it can be shown that an optimum value of the deactivation constant exists for the maximum yield of the intermediate R (Sadana and Doraiswamy, 1971). Since the deactivation constant λ is proportional to the time on stream, it follows that an optimum value of this time also exists for a reaction of this type. However, for this type of consecutive reaction the reaction parameter

B', as defined by equation (8.8), also exhibits an optimum. Hence, it may be concluded that B' and the deactivation parameter, λ, should not be considered in isolation for the optimization of the process time. Prasad and Doraiswamy (1974) showed that the yield of R is a maximum when the product $B'(1 - \lambda)$ is constant; the value was given by:

$$B'(1 - \lambda) = \left(\frac{S}{S - 1}\right) \ln S \qquad (8.54)$$

where S is the selectivity ratio of the two consecutive rate constants, $k_1(0)/k_2(0)$, when no deactivation is occurring. This ratio is determined only by the kinetics of the reaction. It is therefore possible to maintain the yield of intermediate R at the maximum level corresponding to $B'_o = [S/(S - 1) \ln S]$ by manipulating B' when λ changes. This can only be done in the course of reaction by adjusting the feed rate to the reactor so that:

$$B' = \frac{B'_o}{1 - \lambda} \qquad (8.55)$$

This equation is plotted in Fig. 8.7 and shows that, when λ takes values very close to unity, B' tends to infinity and the feed rate tends to zero. Thus the maximum utility of cycle time can be achieved with the minimum number of regenerations by continuously varying the feed rate according to equation (8.55).

This type of consecutive reaction is frequently encountered industrially, and it is of interest to examine the overall optimization of such a reaction

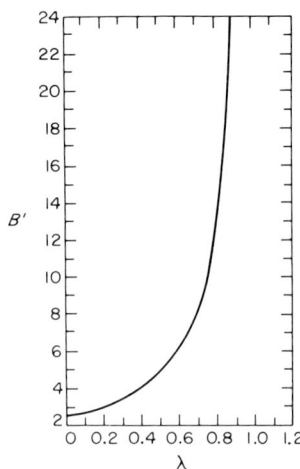

Fig. 8.7 Variation of B' with λ for optimal policy. (Prasad and Doraiswamy, 1974)

8 Optimization of Deactivating Reactor Systems

sequence, taking into consideration the unit costs of the process. To obtain the optimum value of the time on stream, equation (8.55) is written in terms of the feed rate:

$$\bar{F} = F_o(1 - \lambda \bar{t}) \tag{8.56}$$

where F_o and $\bar{F}$ are the initial feed rate and feed rate at any subsequent time respectively and $\bar{t}$ is the time normalized with respect to the time on stream. If it is assumed that the process time is equal to the regeneration time, the total cycle time is $2t_m$ and the frequency $1/2t_m$. If the plant were to operate with the maximum yield of R under all conditions, the rate of production of R at any instant during the cycle is $\bar{F} Y_{R,\max}$ where $Y_{R,\max}$ is the maximum yield of R. Alternatively, using equation (8.56), the total plant capacity can be expressed as:

$$C = \frac{1}{2t_m} \int_0^1 F_o Y_{R,\max} t_m (1 - \lambda \bar{t}) \, d\bar{t} \tag{8.57}$$

$$= \tfrac{1}{2} F_o Y_{R,\max}\left(1 - \frac{\lambda}{2}\right) \tag{8.58}$$

where F_o is a fixed quantity and is given by:

$$F_o = \frac{(S-1)W k_{1,0}}{S \ln S} = \frac{W k_{1,0}}{B'} \tag{8.59}$$

The total capacity at $t_m = 0$ is equal to $\tfrac{1}{2} F_o Y_{R,\max}$ while at $t_m = 1/\alpha$, corresponding to $\lambda = 1$, it is $F_o Y_{R,\max}/4$ (α is the decay rate constant). Hence this gives a quantitative insight into how the plant capacity diminishes with increasing t_m.

If it is assumed that the plant is designed to operate at F_o disregarding any catalyst deactivation, the production cost can be minimized by consideration of the catalyst decay phenomenon. Assuming that the unit cost can be expressed as the sum of two functions $\psi(t_m)$ and $\phi(t_m)$, where $\psi(t_m)$ represents the increase in cost due to reduction of capacity and $\phi(t_m)$ the decrease in cost due to reduced frequency, both of these result from increasing the time on stream. Since $\psi(t_m)$ is an increasing function and $\phi(t_m)$ a decreasing function, an optimum value of t_m must exist which corresponds to the minimum cost. This can be found by differentiating the sum of these two functions and equating to zero:

$$\frac{\partial [\psi(t_m)]}{\partial t_m} + \frac{\partial [\phi(t_m)]}{\partial t_m} = 0 \tag{8.60}$$

Obviously the two cost functions must be known for this relation to be applied, but these can usually be obtained quite readily.

It should be stressed that the optimal feed policy is only valid when the time for the regeneration portion of the overall cycle is independent of the reaction part of the cycle. Thus, if deactivation is caused by coking of the catalyst and the coke deposited is a function of feed rate, the optimal feed policy must account for the burning time in the regeneration part of the cycle. As shown previously, however, in many cases the deposition is only a very weak function of feed rate.

The problem of devising an optimal start-up policy for a deactivating reactor has not been studied to any extent to date. The optimal temperature policies outlined above are basically of two types. In the first, which has been the most thoroughly investigated, the temperature of the reactor is increased in order to make up the loss in catalyst activity so that constant conversion is maintained. In the second, the reactor is operated at the maximum allowable temperature. For both policies the assumption of pseudo-steady state is generally made. However, under start-up conditions this assumption cannot be valid and use of this could lead to misleading results.

Kao (1980) has investigated this problem of start-up for a deactivating catalytic reactor using a stirred tank reactor which is deactivated either by independent poisoning or by series fouling. A singular perturbation analysis was used in the analysis. The optimal policy for start-up for all cases of series deactivation and for some cases of independent deactivation was found to be as follows. The reactor should be started up at the maximum allowable operating temperature for a prescribed period of time before switching to a constant conversion policy typical of the pseudo-steady state analyses. Kao observed that the conversion level selected for the optimal temperature policy under pseudo-steady state conditions depended on the reactor residence time, the operating cost, and the sale price of the product. The improvements obtained are more significant when the operating cost of the reactors is close to the sale price of the product.

References

Andrews, J. M. (1959). *Ind. Eng. Chem.* **51**, 507.
Chou, A., Roy, W. H. and Aris, R. (1967). *Trans. Inst. Chem. Eng.* (London) **45**, T153.
DeLancey, G. B. (1973). *Chem. Eng. Sci.* **28**, 105.
Froment, G. F. and Bischoff, K. B. (1962). *Chem. Eng. Sci.* **17**, 105.
Kao, Y. K. (1980). *Chem. Eng. J.* **20**, 237.
Lee, G. H. and Crowe, C. M. (1970). *Chem. Eng. Sci.* **25**, 743.
Levenspiel, O. (1972). *Chemical Reaction Engineering*, 2nd Edn. John Wiley and Sons.
Miertschin, G. N. and Jackson, R. (1970). *Can. J. Chem. Eng.* **48**, 702.

Ogunye, A. F. and Ray, W. H. (1968). *Trans. Inst. Chem. Eng.* (London) **46**, T225.
Park, J. Y. and Levenspiel, O. (1976). *Ind. Eng. Chem.* (*Proc. Des. Devel.*) **15**, 534.
Prasad, K. B. S. and Doraiswamy, L. K. (1974). *J. Catal.* **32**, 384.
Sadana, A. and Doraiswamy, L. K. (1971). *J. Catal.* **23**, 147.
Sadana, A. and Levenspiel, O. (1978). *J. Catal.* **33**, 1393.
Szépe, S. and Levenspiel, O. (1968). *Chem. Eng. Sci.* **23**, 881.
Voorhies, A., Jr (1945). *Ind. Eng. Chem.* **37**, 318.
Weekman, V. W., Jr (1968). *Ind. Eng. Chem.* (*Proc. Des. Devel.*) **7**, 90.
Young, P. W. and Clark, C. B. (1973). *Chem. Eng. Progr.* **69**, No. 5, 69.

9
Regeneration of Deactivated Catalysts

On previous pages it has been shown that deactivation of catalysts is an important feature in the operation of reactors. After a certain time on stream, which may vary from a few seconds to several years, the activity of the catalyst may be reduced to a level that makes continued operation uneconomic. Additionally, and frequently more important, the selectivity of the catalyst may have been reduced prior to any appreciable loss in activity, so the catalyst performance is inefficient. This happens with, for example, the silver catalysts used in the production of ethylene oxide. When selectivity or activity, or both, fall off, the catalyst charge must either be regenerated or renewed. Whether the catalyst has to be replaced or whether it can be regenerated depends on the type of process responsible for the deactivation. The ease with which a possible regeneration can be carried out depends in turn on the type of reactor as well as the process. Thus regeneration of coked catalysts in a reactor where there is continuous solids transport, such as in fluid or moving beds, is comparatively simple. At the other extreme, in certain fixed bed operations the catalyst may be discarded or the spent catalyst discharged and regenerated elsewhere.

Obviously the arguments as to whether a catalyst should be discarded or regenerated depends to a great extent on economic factors. Chief of these is the cost of catalyst as compared with the useful catalyst life obtained in the reactor. If a catalyst has a life of several years in a fixed bed, and is not of the precious metal type, regeneration may not be worthwhile.

9.1 Feasibility of regeneration

Catalyst deactivation can be due to (a) poisoning, (b) sintering, and (c) fouling. Poisoning of catalysts can be further divided into reversible and irreversible poisoning processes. Reversibly poisoned catalysts represent a comparatively trivial regeneration problem. For most of these, removal of the poison from the feed stream may suffice, but in some cases additional treatment may be required. For example, if a nickel catalyst is deactivated by traces of oxygen in the feed stream, removal of the oxygen impurity from the feed followed by reactivation by reduction in a hydrogen atmosphere at high temperature will usually restore catalyst activity to a level close to the original value.

Irreversible poisoning, as the name implies, is not generally susceptible to regeneration. The classic example of irreversible catalyst poisoning is that by sulphur compounds on reduced nickel catalysts. Even here, however, attempts have been made to determine whether these catalysts can be regenerated. Rostrup-Nielsen (1968) has shown that the adsorption of H_2S on a supported nickel catalyst is in fact reversible, with the coverage being a function of $P(H_2S)/P(H_2)$. However, in spite of this reversibility it is difficult to regenerate these catalysts poisoned by H_2S. Steam has no effect on the chemisorption equilibrium (Rostrup-Nielson, 1971) but steaming at 600–650°C can remove sulphur from unpromoted catalysts by complete oxidation of the nickel. With the usual alkali promoted nickel catalysts, however, steaming converts the sulphur into alkali sulphate. Since the latter are stable, regeneration of this type of catalyst is difficult. The situation can probably be best summed up as conceding that, while these catalysts may be regenerated with some difficulty in laboratory experiments, in practice the sulphur poisoning is essentially irreversible.

Sintering by its very nature tends to be an irreversible process, although some degree of redispersion of an aged supported platinum catalyst may be obtained by partial oxidation during operation. At present, however, such methods are not too well understood, and therefore for most practical applications the catalyst is regarded as non-regenerable and is usually discarded for platinum recovery and reformulation at the end of its active life.

As already pointed out, fouling may be of two main kinds. One is the deposition of carbonaceous material or "coke" on the catalyst during the processing of organic feedstocks, while the other is the deposition of metals from petroleum feedstocks resulting in pore plugging. In the first case the catalyst may be taken off stream and heated to a moderate temperature in an atmosphere containing some oxygen so that oxidation or "burn-off" of the coke is achieved. Sometimes it may first be necessary to treat the catalyst with steam in order to remove liquid organic products adhering to the catalyst. The main problem in the regeneration of coked catalyst particles is the minimization of the temperature rise caused by the exothermic oxidation reaction when the coke is converted into CO_2 and CO. This can cause thermal sintering of the catalyst unless care is taken; the problem is greatest when fixed beds are employed because of the difficulty of adequate heat removal in reactors of this type. The usual procedure is to admit only a very low concentration of oxygen into the reactor in the early stages of regeneration; with increasing time the oxygen concentration is increased until complete oxidation of the coke deposit is achieved.

When fouling occurs by metal deposition, as for example the vanadium or nickel deposits, in the hydrotreating of some petroleum feedstocks and the iron and nickel deposits in the treating of coal derived liquids, the simple oxidation process described above is not applicable. In these circumstances the catalyst has to be discharged from the reactor and the material reprocessed.

Thus the only current deactivation process that lends itself to facile catalyst regeneration is that which occurs when coke deposits are formed on the active catalyst surface. Since this is one of the main areas of catalyst deactivation, extensive studies have been undertaken to determine the optimum conditions for regeneration. The remainder of this chapter is devoted to this important problem.

9.2 Description of coke deposit and kinetics of regeneration

A large amount of published literature exists on the kinetics of coke burn-off from a variety of catalysts. An appreciable amount of this literature is concentrated on silica–alumina and the related zeolite catalysts owing to the industrial significance of cracking operations.

Microscopic examination of the coked catalyst has shown (Haldeman and Botty, 1959) that the carbon phase is normally well dispersed within the granules and there is no preferential internal deposition of carbon. Most of the coke consisted of thin filmy aggregates of particles less than 10 nm in size. Massoth (1967) also found that particles of coke were of this size when pellets were coked at 375°C, while Hughes and Shettigar (1971) deduced from kinetic results that coke particles should be about 4 nm in diameter. Calculations also show that the particles do not (for coke concentrations of 1–5%) spread out to cover all the available catalyst surface but tend to cluster as discrete assemblages on the surface; this has been confirmed by microscopic examination.

X-ray diffraction studies show that approximately 50% of the carbon deposit exists as pesudo-graphitic structures. The remainder probably consists of unorganized aromatic systems and of aliphatic appendages to polynuclear aromatic systems. It is generally agreed by all investigators that the molecular formula of the coke deposit varies between $C_1H_{0.4}$ and C_1H_1. This demonstrates that significant quantities of hydrogen are present, so the oxidation will produce not just CO_2 and CO but appreciable quantities of water vapour as well. This important feature will be discussed in detail later. Although Weisz and Goodwin (1966) found that burn-off of coke was remarkably independent of the nature and method of deposition, they may not have noted the effects of different hydrogen concentrations since their studies were concentrated on the overall combustion behaviour.

Walker *et al.* (1959) extensively reviewed the literature on the products of carbon combustion, and have drawn the following conclusions.

(a) Both carbon monoxide and carbon dioxide are the primary products of carbon combustion. The oxidation of carbon to carbon monoxide and carbon dioxide is not significantly restricted by equilibrium considerations even at temperatures of 4000 K.
(b) The ratio CO/CO_2 is relatively independent of the type of carbon used. The empirical relations used by various workers to predict this ratio give values between 0.3 and 0.9.
(c) The primary CO/CO_2 ratio increases with reaction temperature.
(d) Lower gas velocities tend to increase the CO_2 content because of increased secondary oxidation of CO to CO_2. The presence of water vapour also increases this secondary oxidation.

The three rate controlling steps generally encountered in the burn-off of carbon from coked catalysts are (1) control by intrinsic carbon burning kinetics, (2) control by oxygen diffusion through the catalyst pores, and (3) control by both (1) and (2).

In the kinetic controlled region of "intrinsic" chemical kinetics, combustion is controlled by the intrinsic burning rate of carbon. The reacting oxygen diffuses to all coke particles throughout the pellet, and consequently combustion occurs throughout the pellet but at different rates at different radial positions. The mass transport or diffusion controlled region is characterized by the "shell progressive" mechanism already referred to. The reaction zone in this process moves as a thin shell from the outer surface towards the centre as reaction proceeds. The diffusion rate of oxygen, being significantly smaller than the reaction rate of coke, controls the combustion reaction. In general, however, the rate of regeneration is not wholly controlled by either mechanism since both steps are operating with varying degrees of importance. When both steps have approximately equal importance the burn-off is said to be under intermediate control.

Increase of temperature results in a transition from the kinetic controlled region, through the intermediate control region to the mass transport dominated region. Weisz and Goodwin (1963) demonstrated this pictorially in the form of concentration profiles of coke for the three regions, as shown in Fig. 9.1. For the case of silica–alumina beads of 4 mm diameter, kinetic control was observed up to about 475°C and mass transport control took hold at temperatures above 625°C.

Weisz and Goodwin also investigated the effect of particle size by

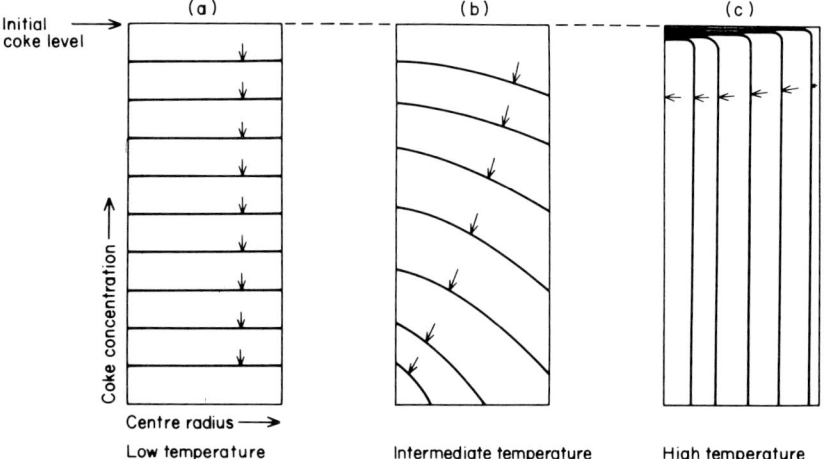

Fig. 9.1 Coke concentration profiles in catalyst beads for successive stages of burn-off. (Weisz and Goodwin, 1963)

considering beads of 4 mm diameter and powdered catalyst of 0.2 mm diameter. The results obtained are shown in Fig. 9.2 where both sizes of materials have their oxidation rates plotted against temperature. The straight line for the powder demonstrates that the process was kinetically controlled throughout the temperature range. The beads, however, gave a transition to diffusional control at higher temperatures.

The intrinsic carbon burning rate is the rate of burn-off of coke in the

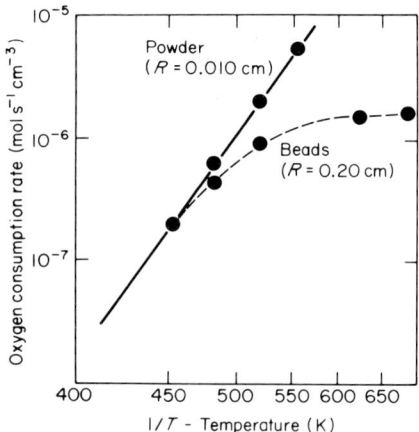

Fig. 9.2 Average observed burning rates of silica–alumina cracking catalyst (C_{co} = 3.4%). (Weisz and Goodwin, 1963)

absence of appreciable diffusional restrictions, and it has generally been observed to be proportional to oxygen concentration, but the dependence of the reaction rate on the carbon concentration shows some variation according to the individual investigations. One possible reason for this variation is that in some cases there may have been an ageing effect causing a decrease in carbon activity (Johnson and Mayland, 1955). Massoth (1967) claimed that the reaction was not first order in carbon but was surface reaction controlled. Hughes and Shettigar (1971), using a statistical analysis of their rate data, concluded that the surface reaction expression gave a better fit than the first order process. The reaction of the hydrogen constituent of the coke appears not to have been investigated in detail. Massoth (1967) observed that the concentration of water vapour was initially very high, decreasing rapidly however as reaction proceeded. Consequently, Massoth suggested that the initial temperature peak observed could be attributed largely to the hydrogen content of the coke. He therefore proposed a model in which the hydrogen in the coke was oxidized rapidly leaving a residue of hydrogen free carbon which reacted at a lower rate. A model was proposed but Massoth was not able to fit this to his results. Subsequent work by Ramachandran *et al.* (1975) produced a model that could interpret some experimental results (see below).

The activation energies reported for the oxidation of coked catalyst particles exhibit a wide variation as Table 9.1 shows. The very low values are probably caused by diffusional restrictions and are not true activation energies as such. It seems that a value of 150 kJ mol^{-1} is representative of the rate coefficient for the oxidation of carbon.

Parvinian (1977) measured CO_2 and CO as well as H_2O evolution as a function of time during the oxidation of coke from a silica–alumina bead catalyst. He found that production of CO_2, CO, and H_2O were all first order in oxygen concentration and production of total carbon oxides was first order in carbon. The activation energy for water vapour production was

Table 9.1 Activation energy of the regeneration reaction.

Temp. range (°C)	E_A (kJ mol^{-1})	Reference
⩽480	167	Massoth (1967)
420–544	134	Hughes and Shettigar (1971)
450–600	157	Weisz and Goodwin (1966)
450–600	171	Johnson and Mayland (1955)
455–600	111	Dart *et al.* (1949)
600–900	77	Mickley *et al.* (1965)
430–550	151	Parvinian (1977)

75 kJ mol^{-1} and confirmed Massoth's assumption that hydrogen was removed more readily than carbon in the overall coke oxidation process.

9.3 Regeneration of fluidized bed catalysts

Since a fluid bed cracker contains particles of small size (typically less than 100 μm in diameter on average) diffusional restrictions within the particles are absent and therefore these systems are simpler to analyse than pelleted catalysts. The catalysts used are now generally synthetic zeolites although some of the less reactive silica–alumina powders may be still used for some applications. Bondi *et al.* (1962) showed that even for these small particles considerable temperature rises (to nearly 900°C) within the particles could occur if unburnt oil was carried over into the regenerator by the particles.

If steady state operation is required the cracking reaction rate (and hence the coke formation rate) cannot increase beyond the rate of coke oxidation in the regenerator (Venuto and Habib, 1979). With the development of highly active cracking catalysts, such as zeolites, this has posed severe problems in obtaining an efficient regenerator design and operation. Additionally, it is in the aggressive environment of the regenerator that most damage to the catalyst can occur. The primary objective of regeneration is to obtain a low residual level of coke on the catalyst but at a high burning rate so that the residence time is not excessive and thereby avoiding high temperature deactivation of the catalyst.

"Make-up" fresh catalyst is added periodically while at the same time the aged catalyst is withdrawn to maintain the activity at an appropriate average level in the unit. It is desirable to have a small catalyst inventory, consistent with regeneration kinetics, producing a regenerated catalyst containing about 0.1–0.2% carbon (the coke content resulting from cracking is usually about 1–2%). A smaller inventory of catalyst allows a faster response in the system to catalyst make-up and requires less of the costly fresh catalyst to maintain a given equilibrium activity. Also less attrition occurs under conditions of small catalyst inventories.

Good distribution of combustion air and efficient fluid gas contacting are crucial in the regenerator. This is because of the requirement to obtain as uniform an oxidation rate as possible and avoid the problem of localized high temperatures which could occur if this were not done. Usually adequate mixing can be provided by the use of suitable mechanical design features in the regenerator to promote mixing. This is also becoming more important with the trend towards higher regenerator temperatures. This, in turn, has resulted from (a) the necessity to obtain lower carbon levels on the regenerated catalyst to promote improved accessibility to the zeolitic structure, (b) the higher operating temperatures of most crackers at present, and (c) because of the desirability of converting more CO into CO_2 in the regenerator rather than to have this reaction later in the process (after-burning).

The main factor influencing the coke oxidation rate is the oxygen partial pressure. Ideally, oxygen levels in the flue gas from the regenerator should be less than 1% so that maximum utilization of the oxygen could be achieved while minimizing possible problems of after-burning of CO to CO_2. With satisfactory distribution of air the number of kg of air required to burn 1 kg of coke is constant at about 11–14. The CO/CO_2 ratio in the flue gas has been found to depend on the mode of regeneration and the type of catalyst used. Typically, a value of 1 has been obtained in many cases. After-burning, that is the uncontrolled combustion of CO to CO_2 that can occur in the dilute phase of the regenerator if excess oxygen is present, can cause severe operating problems. Runaway temperatures exceeding 760°C have been reported (Murphy and Soudek, 1977) when after-burning occurs. Frequently after-burning arises because of the presence of metal contaminants in the catalyst and iron particles carried over from the regeneration zone proper.

It would seem that, although problems can arise in fluid bed systems, the flexibility of operation of those units plus the excellent heat transfer characteristics of fluidized beds have enabled most problems to be overcome even under the most exacting operating conditions.

With fixed beds there are a number of problems and these will be considered next. The analysis will first be developed for single catalyst pellets and this will be followed by a discussion of reactor performance.

9.4 Regeneration of coked catalyst pellets

The regeneration of a catalyst pellet by burning off the carbonaceous deposit of coke in the pores represents a special case of non-catalytic gas–solid reactions. There is a wealth of literature on non-catalytic processes, reflecting their industrial importance, since such processes include the reduction of iron ore, calcination of limestone, combustion of solid fuels, etc. The regeneration process is a special case of combustion in which the fuel (the coke) is deposited within a porous matrix by the reaction process and is subsequently removed from the matrix by oxidation to combustion gases. Thus the catalyst pore structure exists as a reservoir into which coke is deposited and removed, and in ideal operation the pore structure should remain invariant. In this respect it differs from reactions such as the combustion of a lump of coal where the pore structure is constantly changing during combustion.

Another difference from a number of other non-catalytic processes is that in many of these an isothermal analysis will suffice since, although the reaction may not be isothermal, the conversion may be predicted sufficiently accurately for design purposes from any isothermal model. When catalysts are being regenerated by burning off the coke, in addition to the conversion rate (time required for regeneration) it is important to establish the maximum temperature during the process in order to guard against catalyst sintering. Thus in catalyst regeneration a non-isothermal exercise has usually to be followed if all operating requirements are to be fulfilled.

In general, the mechanism of regeneration of single catalyst pellets depends very much on the temperature range used as depicted in Fig. 9.1. At low temperatures oxygen is accessible to all the coke deposits throughout the particle, and a homogeneous type of gas–solid reaction occurs similar to that in a normal catalytic reaction with a value of the effectiveness factor close to unity. At the other extreme, at high temperatures, the oxidation rate becomes very fast, so the reaction becomes limited by mass transport of oxygen through the pores. Under these conditions a shell progressive type mechanism operates as shown in Fig. 9.1(c).

Early work on pellets was concerned mainly with the isothermal problem and in estimating the time required to remove all coke from the pellet. Ausman and Watson (1962) identified two distinct rate periods called the constant rate and falling rate periods. The former was identified with burn-off of the surface coke, and Ausman and Watson obtained oxygen concentration profiles for this period. The falling rate period is related to the burn-off of coke within the pellet, and oxygen profiles were obtained for this case also, but the rate expression for this period is a complex function of the fraction of oxygen remaining. Dobychin and Klibanova (1959) also proposed a two stage model based on a similar distribution of coke deposits between the surface and the interior. In fact, however, both sets of authors have used a rather complicated two-stage mechanism to explain results which could be explained more simply in terms of a general diffusion controlled model for coke burn-off, and the elaboration required in suggesting a change in mechanism from "surface" reaction to "interior" reaction is not justified.

Richardson (1972) has obtained direct evidence of a shell progressive mechanism in his studies of carbon profiles on cobalt molybdate catalysts where a combustion technique was used to measure the CO_2 liberated. However, probably the most detailed experimental investigation of conversion and especially of temperature profiles in catalyst pellets during coke burn-off was made by Shettigar and Hughes (1972b). In this work temperature profiles in coked catalyst pellets undergoing regeneration were measured under transient conditions for a number of instrumental pellets containing Pt/Pt–13%Rh thermocouple wires of 0.025 mm diameter.

9 Regeneration of Deactivated Catalysts

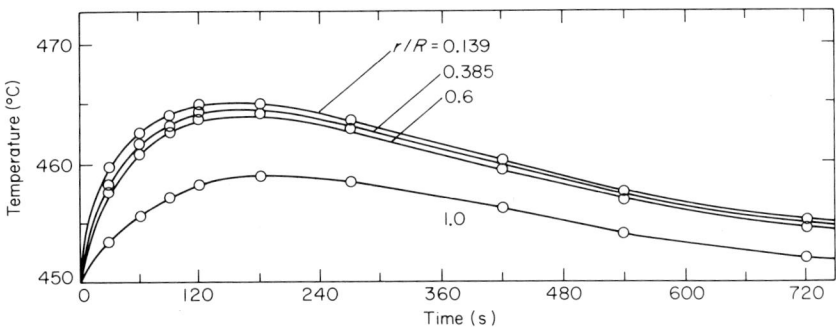

Fig. 9.3 Measured transient temperature profiles at 450°C. (Shettigar and Hughes, 1972b)

Different pellets contained thermocouples with junctions at different positions, thus enabling a reasonably comprehensive picture of the variation of temperature with position and time to be obtained. Pellets were of 12.5 mm diameter and were coked using xylene. The development of temperature profiles during reaction was measured at four radial positions in the pellet for temperatures in the range 450–520°C at a constant gas flow rate and an initial coke deposit of 1% by weight. The results are shown in Figs 9.3 and 9.4.

Figure 9.3 shows a typical measured transient temperature profile for the combustion of coke at an initial temperature of 450°C. At this temperature the oxidation would be expected to occur homogeneously throughout the pellet. The profiles followed the same pattern for each radial position, showing a maximum value in the initial stages of reaction, followed by a more gradual decrease. At any time the temperature increased with position towards the centre of the pellet although this difference was small except for positions $r/R = 1$ to $r/R = 0.6$. The maximum temperature rise recorded was 15°C and verification that the reaction was indeed homogeneous was obtained by sectioning a pellet partially reacted under identical conditions; a uniform deposition was found. When the temperature was raised to 500°C or higher, however, sectioning showed that two zones developed, an outer completely reacted shell and an inner unreacted cone. Temperature profiles were obtained in this region also, and Fig. 9.4 shows these for an initial temperature of 550°C. The reaction zone was now established at the exterior surface of the pellet, and the temperature of this zone increased at a much faster rate than at any other radial position. With progression of the reaction zone towards the centre of the pellet and the forward removal of heat from the reaction zone, the interior temperature reached a maximum value and the temperatures in the outer regions decreased gradually (Fig. 9.4). At this stage

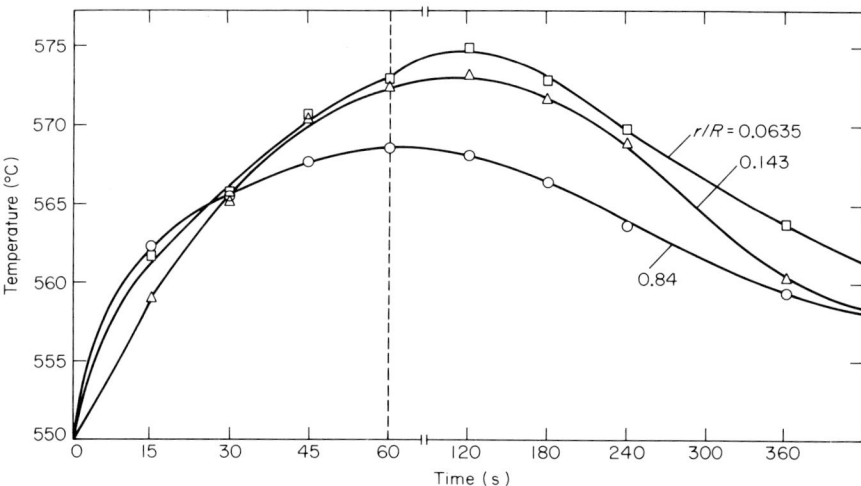

Fig. 9.4 Measured transient temperature profiles at $T_o = 550°C$; coke level = 1.0 wt%; air flow rate = 1.67×10^{-5} m^3 s^{-1}. (Shettigar and Hughes, 1972b)

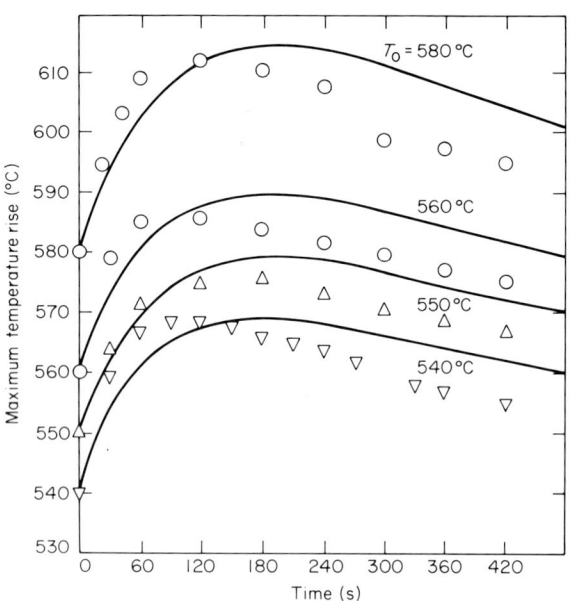

Fig. 9.5 Comparison of maximum measured temperature rise at different initial temperatures with predictions of finite thickness; reaction zone model. (Shettigar and Hughes, 1972b)

the reaction rate had decreased considerably, so the rate of heat loss was greater than the rate of heat generation. Consequently, the temperature of the pellet as a whole started to decrease even before the completion of the reaction. The maximum temperature rise at 550°C was about 25°C and occurred close to the central region of the pellet.

The variation of the maximum temperature rise during regeneration with initial temperature was also determined by Shettigar. Here maximum temperature rise means the outer envelope of the temperature curves for different radial positions as a function of time, as plotted in Figs 9.3 and 9.4. It was found that this maximum temperature rise was strongly dependent on the initial temperature, as shown in Fig. 9.5, an initial temperature of 580°C giving a temperature rise of over 30°C for a 1% weight of coke on the catalyst.

The transient temperature profiles were also very sensitive to the amount of coke deposited on the catalyst. Figure 9.6 shows the maximum temperature rise plotted against time for coke levels of 1.0% and 1.8% carbon. The

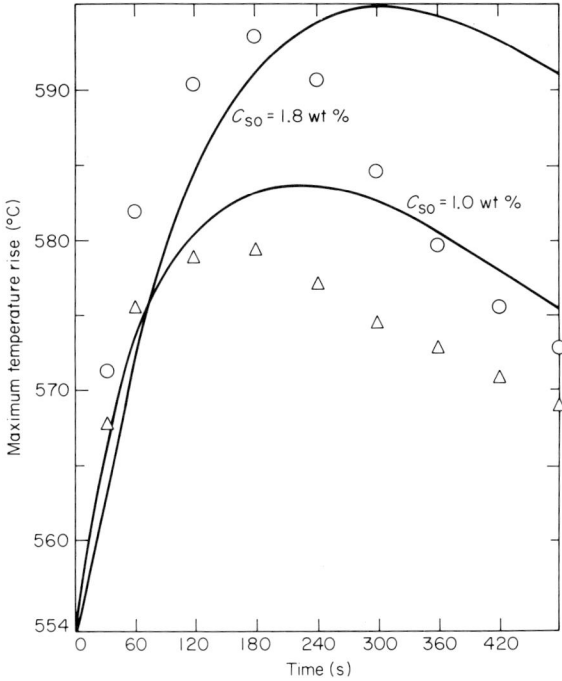

Fig. 9.6 Comparison of measured maximum temperature rise at two different coke concentrations with the predictions of the finite thickness reaction zone model. (Shettigar and Hughes, 1972b)

appropriate maximum temperature rises attained were 26 and 40°C respectively at an initial temperature of 544°C.

Many investigators have derived theoretical models for the conversion of non-catalytic gas–solid reactions. In general it is not difficult to obtain approximate agreement between experiment and model when conversion is the basis for comparison. A far more rigorous test is to compare the temperatures predicted by the model as well as the conversion. Unfortunately this has seldom been done, and the reliability of the various models to represent coke burn-off from catalysts has therefore not always been unequivocally demonstrated.

Models for coke oxidation in single catalyst pellets can be split into two general groups. These are the homogeneous models and retracting core models. In all, a number of models have been proposed, including some that allow for cracking, shrinkage, etc., but the split into two groups is still generally valid. Some models combine the basic features of each group, and the particle pellet (or grain) model is one such example. The models discussed below include: (i) the homogeneous model; (ii) the sharp interface retracting core model; (iii) the finite thickness reaction zone, retracting core model; (iv) the particle pellet model; (v) various specific models for coke oxidation which include some or all of the features of the above models.

When chemical kinetics is the rate controlling step, gaseous reactants will diffuse throughout the particle and the reaction may be visualized as occurring throughout the solid phase, although the actual reaction rates at different positions will vary. This kind of homogeneous reaction forms the basis for the mathematical model of the same name. Figure 9.7(a) shows the change in the concentration profile of the solid with radius after a particular time from the commencement of the reaction. Several workers have contributed to the mathematical development of this model.

With increase of temperature, an exponential increase of the chemical rate will occur. When this becomes sufficiently large, diffusion of the gaseous reactant into the solid pores will assume the role of the rate controlling step. This situation may also come about without significant increase of temperature if the solid is relatively non-porous. Under these circumstances the gaseous reactant will be immediately consumed at the reaction zone. The reaction zone itself may be visualized in this case to consist of a sharp interface travelling from the outer solid surface towards the centre with depletion of the solid reactant. Figure 9.7(b) shows the variation of the solid concentration with radial position after a particular time from the commencement of the reaction in a spherical particle. Yagi and Kunii (1961) were the first to postulate the mechanism for this sharp interface unreacted shrinking core model. Many other workers have also studied various aspects of this model.

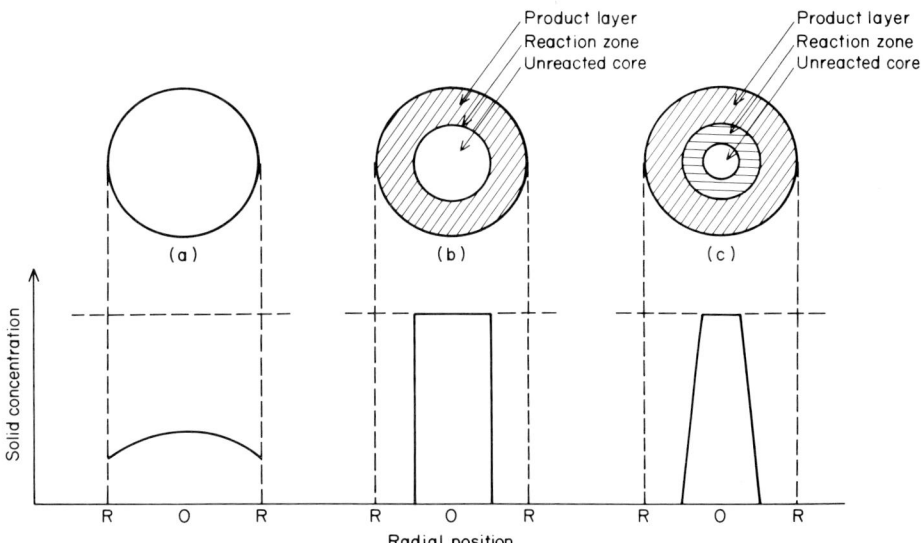

Fig. 9.7 Diagrammatic representation of various single pellet models.

As has been mentioned earlier, total control of the reaction rate by either intrinsic kinetics or diffusional characteristics of the system represents only extreme cases. The more general situation is where both of these steps play a significant role in the overall rate control. Ausman and Watson (1962) were the first to recognize this in their model. However, the general model postulated by them and later followed by Ishida and Wen (1968) had a serious shortcoming in that it assumed the order of the reaction with respect to the solid reactant to be zero. The more general case is where the reaction is first order with respect to both the solid and the gas.

Bowen and Cheng (1969) modified the sharp interface unreacted shrinking core model to suit the intermediate region of reaction control. They suggested that it is more realistic to assume a narrow but finite thickness reaction zone between the product layer and unreacted core instead of one with a sharp interface. This in effect means that the diffusing gas does not get consumed immediately on contact with the reaction zone but only after penetration of a small but finite length into the unreacted core. Reaction occurs all over this finite length. Under extreme conditions of total kinetic or diffusion control, this reaction zone either extends all over the particle or contracts to give a sharp interface. Further development of the finite thickness reaction zone unreacted shrinking core model was carried out by Shettigar and Hughes (1972a). Figure 9.7(c) gives the variation of solid concentration with radius after the reaction has progressed for some time, in the case of this model.

In practice, a porous solid often comprises an agglomeration of smaller particles, and this particulate nature of the solid is recognized in the particle pellet or grain model (Calvelo and Smith, 1970; Szekely and Evans, 1971). Initially, simplifying assumptions were introduced. Thus, Szekely and Evans assume isothermality in the pellet and neglect ash layer diffusion, while Calvelo and Smith neglect heat and mass transfer resistances. A comprehensive development of this model, including all resistances and with a transient heat balance, was given by Sampath *et al.* (1975a). The model is depicted in Fig. 9.8. Basically, the model assumes that reaction occurs within the particles making up the pellet by a sharp interface mechanism. Transport to the particles is via the interparticle voids within the pellet, and this diffusional restriction in the macropores causes variation in the rates of reaction

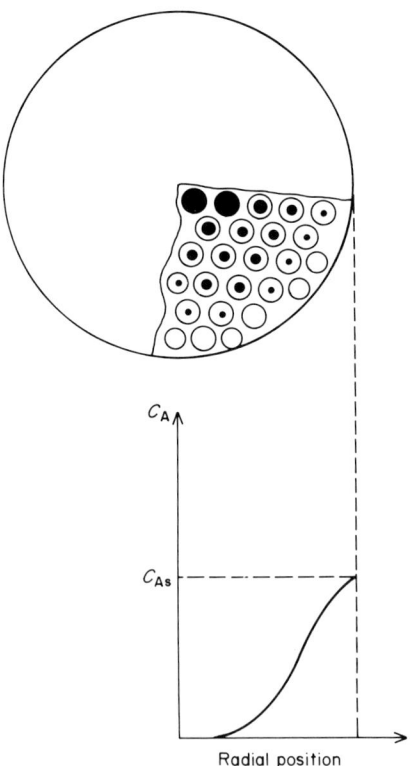

Fig. 9.8 The particle pellet model.

9 Regeneration of Deactivated Catalysts

of the individual particles according to their accessibility to the reactant gas. Because the particles are small, isothermality may be assumed within the particles.

For all retracting core models the general assumption is made that, relative to gaseous reactant diffusion in the product layer, the unreacted core is stationary. This vastly simplifies the solution of the non-steady state diffusion problem with a moving boundary. The assumption of a pseudo-steady state is justified because of the much greater rate of transport of the gaseous reactant towards the unreacted core compared with the rate of movement of the core itself. Bischoff (1963) has critically examined this assumption and found that it was of general validity for this type of non-catalytic gas–solid reaction. Thus, in practice, the accumulation term may generally be neglected in the mass balance.

However, for the heat balance, the accumulation term is often comparable in magnitude to the heat of reaction. Therefore use of the steady state heat balance would lead to erroneous results. Full transient heat balances were employed by Bondi *et al.* (1962), Luss and Amundson (1969), Shettigar and Hughes (1972a), and Sampath *et al.* (1975a, b), following the suggestion made by Beveridge and Goldie (1975).

The development of these various models is given in the original papers. The results are summarized below.

Mass and heat balances for all four models are presented in Table 9.2 (mass balances) and Table 9.3 (heat balances). In these tables, C, ρ, and T represent the dimensionless concentration, reaction radius, and temperature respectively, and ϕ, ϕ_o, and ϕ_a are Thiele moduli as defined below. Also, for the sharp interface model:

$$F_1 = \phi\left[\left(\frac{1}{\text{Sh}^*} - 1\right)\rho^2 + \rho\right] \tag{9.10}$$

with

$$\phi = k_c R / D_e \tag{9.11}$$

while for the finite thickness reaction zone model:

$$F_1 = \phi_o\left[\left(\frac{1}{\text{Sh}^*} - 1\right)\rho^2 + \rho\right] \tag{9.12}$$

with

$$\phi_o = R\left(\frac{k_c C_{Bo} \sigma}{6n D_e}\right)^{\frac{1}{2}} \tag{9.13}$$

Table 9.2 Solutions for gas phase mass balance equations.

Model	Equation	
Homogeneous	$\dfrac{d^2C}{d\delta^2} + \dfrac{2}{\delta}\dfrac{dC}{d\delta} = R\phi \exp\left[\dfrac{\gamma\beta(u-c)}{1+\beta(u-c)}\right]$	(9.1)
Sharp interface retracting core model	$1 - C - F_1 C \exp\left[\gamma\left(1 - \dfrac{1}{T}\right)\right] = 0$	(9.2)
Finite thickness reaction zone, retracting core model	$1 - C - F_1' C \exp\left[\dfrac{\gamma}{2}\left(1 - \dfrac{1}{T}\right)\right] = 0$	(9.3)
Particle pellet (grain) model	$\dfrac{\partial^2 C}{d\delta^2} + \dfrac{2}{\delta}\dfrac{\partial C}{\partial\delta} - \phi_1^2 \rho^2 C F_2 = N_1 \dfrac{\partial C}{\partial\theta}$	(9.4)
	$\dfrac{d\rho}{d\theta} = -CF_2$	(9.5)

Table 9.3 Transient heat balances.

Model	Equation	
Homogeneous	$\dfrac{\partial^2 T}{\partial \delta^2} + \dfrac{2}{\delta}\dfrac{dT}{d\delta} = R\phi(T - 1 - \beta u)\exp\left[\gamma\left(1 - \dfrac{1}{T}\right)\right] + \dfrac{\partial T}{\partial \theta}N$	(9.6)
Sharp interface retracting core model	$T^* = \dfrac{2}{3}\sum\limits_{n=1}^{\infty}\dfrac{\sin\beta_n\delta}{\delta}f(\beta_n)\int_{\rho}^{1}\rho^*\sin\beta_n\cdot\rho^*\exp\left[-\dfrac{A\beta_n^2}{3\mathrm{Nu}^*}F\right]d\rho^*$	(9.7)
Finite thickness reaction zone, retracting core model	$T^* = \dfrac{2}{3}A\sum\limits_{n=1}^{\infty}\dfrac{\sin\beta_n\delta}{\delta}f(\beta_n)\int_{\rho}^{1}\rho^*\sin\beta_n\cdot\rho^*\exp\left[-\dfrac{A\beta_n^2}{3\mathrm{Nu}^*}F'\right]d\rho^*$	(9.8)
Particle pellet (grain) model	$\dfrac{\partial^2 T}{\partial \delta^2} + \dfrac{2}{\delta}\dfrac{\partial T}{\partial \delta} + \beta\phi_1^2\rho^2 CF_2 = N_2\dfrac{\partial T}{\partial \theta}$	(9.9)

$$F = \dfrac{\rho^* - \rho + \tfrac{1}{2}(\rho^{*2} - \rho^2) + \tfrac{1}{3}\left(\dfrac{1}{\mathrm{Sh}^*} - 1\right)(\rho^{*3} - \rho^3)}{\phi\exp\left[\left(1 - \dfrac{1}{T}\right)\right]}$$

$$F = \dfrac{\rho^* - \rho + \tfrac{1}{2}(\rho^{*2} - \rho^2) + \tfrac{1}{3}\left(\dfrac{1}{\mathrm{Sh}^*} - 1\right)(\rho^{*3} - \rho^3)}{\phi_o\exp\left[\dfrac{\gamma}{2}\left(1 - \dfrac{1}{T}\right)\right]}$$

(ρ = radius of reacting interface)

The function u occurring in the mass balance for the homogeneous model is given by:

$$u = 1 + \frac{\phi_a}{3}\left(\frac{1}{Nu^*} - \frac{1}{Sh^*}\right) \qquad (9.14)$$

where

$$\phi_a = -\frac{3Sh^*}{C_{Ag}}(C_{Ag} - C_{As}) \qquad (9.15)$$

For the particle pellet model:

$$F_2 = \exp[\gamma(1 - 1/T)] \Big/ \left\{1 + \frac{\rho}{Bi}(1-\rho)\exp[\gamma(1-1/T)]\right\} \qquad (9.16)$$

and

$$\phi_1 = R\left[\frac{k_o 3(1-\varepsilon)}{D_e r_o}\right]^{\frac{1}{2}}$$

The main difference in the solutions for the sharp interface and finite thickness reaction zone models is in the employment of a different Thiele modulus for each. The modulus ϕ_o, being a square-root function of the ratio k_g/D_e, will always be smaller than ϕ, and thus estimated values of ϕ will be smaller at the reaction interface for the sharp interface model as compared with the finite thickness reaction zone model.

The transient equations in Table 9.3 do not consider the radiation heat transfer contribution in the energy balance. This term may generally be neglected when the surfaces that radiate to the particle are also at the gas phase temperature T_g; this is usually true for the homogeneous reaction model. However, in the case of the other models, especially those with a retracting interface in the pellet, the pellet surface temperature may be expected to cool considerably as the reaction zone penetrates deep into the pellet and the radiation heat transfer term may become important.

A simplified treatment of this problem can be made if it is assumed that the pellet is isothermal. The energy balance may then be written as:

$$\frac{dT}{d\theta} = H_1 \rho^2 \left(-\frac{d\rho}{d\theta_1}\right) - H_2(T-1) - H_3(T^4 - 1) \qquad (9.17)$$

where the dimensionless time is defined as:

$$\frac{t D_e C_{Ag}}{C_{Bo} R^2}$$

9 Regeneration of Deactivated Catalysts

and

$$H_1 = \frac{3(-\Delta H)C_{Bo}}{\rho_s c_p T_g}$$

$$H_2 = \frac{3kC_{Bo}R}{\rho_s c_p D_e C_{Ag}}$$

$$H_3 = \frac{3\sigma\varepsilon C_{Bo}RT_g^3}{\rho_s c_p D_e C_{Ag}}$$

Equation (9.17), solved with the equation for the rate of retraction of the interface into the pellet ($d\rho/d\theta$), yields the transient temperature profile.

The profiles generated using the sharp interface, finite thickness reaction zone and particle pellet models were compared with the experimentally observed maximum temperatures obtained by Shettigar and Hughes (1972b) for a bulk gas temperature of 580°C (Fig. 9.9). During the initial stages of regeneration, the sharp interface model follows the experimental curve most closely, but the peak temperature rise predicted by this model is significantly

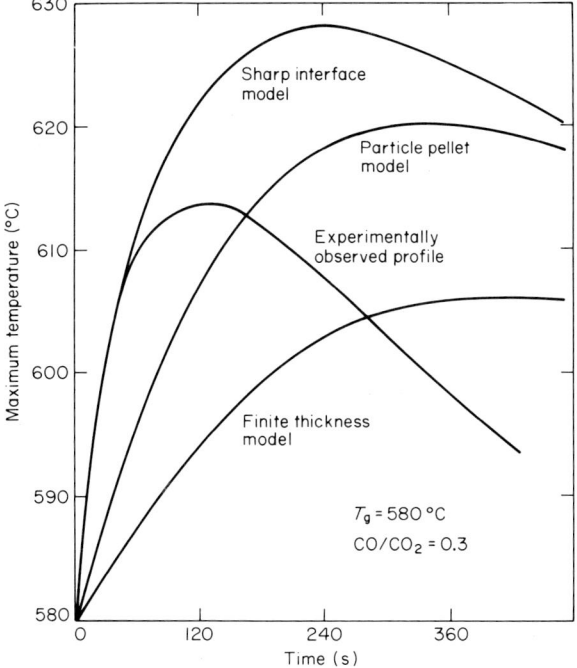

Fig. 9.9 Comparison of maximum temperature profiles at the bulk gas temperature of 580°C. (Sampath et al., 1975c)

in excess of the experimental peak. Although the peak temperature rise predicted by the particle pellet model is in better agreement with the experimental maximum, this occurs at a later time. This displacement also exists for the other two models, with the sharp interface model showing the smallest deviation. None of the models seems able to predict satisfactorily the fairly sharp drop observed experimentally during the latter part of the regeneration, although here again the sharp interface model gives better agreement. One possible reason for this is the non-inclusion of radiational heat transfer between the pellet and the surrounding gas in these simpler models.

The theoretical profiles predicted by the various models assumed that the CO/CO_2 ratio in the product gas was 0.3. The effect of differing values of this ratio is shown in Fig. 9.10 using the sharp interface model at a bulk gas temperature of 580°C (this model is expected to be of greater validity at this

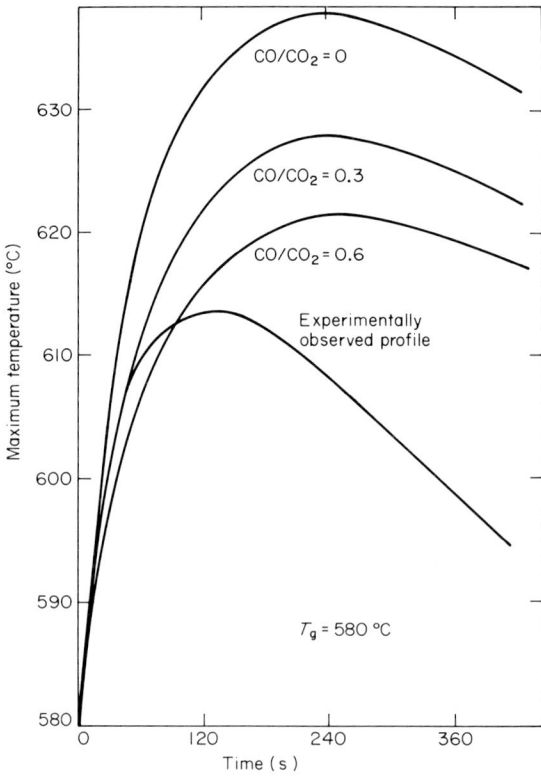

Fig. 9.10 Effect of product gas composition on the maximum temperature profile. (Sampath et al., 1975c)

high reaction temperature). Inspection of this diagram shows that increasing the CO/CO_2 ratio brings the predicted profile into closer correspondence with experiment. This demonstrates the need for precise experimental analysis of all product gases when regeneration is carried out.

Conceptually, the sharp interface retracting core model is best suited to explain the regeneration characteristics at high temperatures when diffusion should dominate the overall reaction process. The finite thickness moving reaction zone model and the particle pellet model would be more suitable for the intermediate control region where both kinetics and diffusion play a role. For regeneration at lower temperatures the homogeneous model would be more applicable. To test this concept the maximum temperature profile as predicted by the various models is compared with experimental results obtained at 450°C (Fig. 9.11). In the computations for this regeneration temperature it was assumed that no carbon monoxide was present as indicated by Walker *et al.* (1959). For all models the agreement with experiment is poor. This is particularly true for the shape of the profiles obtained, where the initial experimental maximum is not reproduced by any model. The maximum temperature attained is approximated most closely by the homogeneous model, but the sharp interface model gives worst agreement at this temperature as might be expected.

One assumption made for all three models is that radiational heat losses between the catalyst pellet and the surrounding gas are negligible. A simple radiation model based on the sharp interface concept, and which assumed

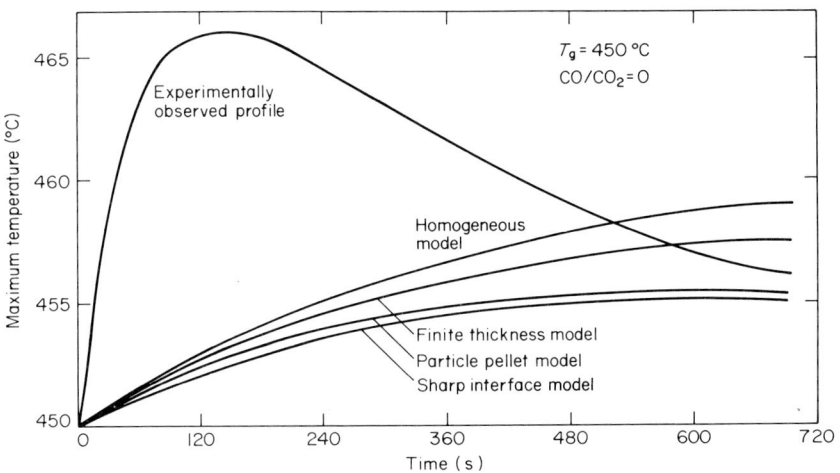

Fig. 9.11 Comparison of maximum temperature profiles at the bulk gas temperature of 450°C. (Sampath *et al.*, 1975c)

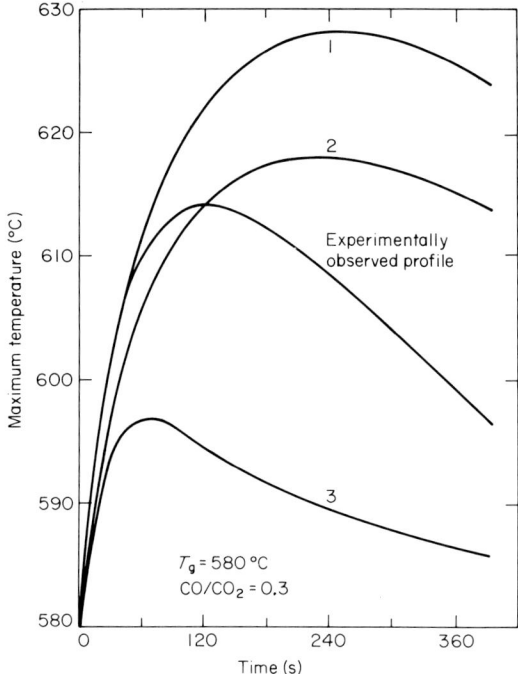

Fig. 9.12 Effect of inclusion of radiational heat losses for sharp interface model. (Sampath *et al.*, 1975c)

the pellet to be isothermal but at a temperature above the surroundings, was therefore formulated to determine whether the influence of radiation was important. This model has already been briefly referred to above; equation (9.17). The maximum temperature profile obtained using equation (9.17) for the heat balance is shown as curve 3 in Fig. 9.12 which compares various models with the experimental maximum temperatures obtained at 580°C. As can be seen, the inclusion of a radiation term causes an appreciable reduction in the maximum temperature attained from curve 1, which is the sharp interface model with internal temperature gradients but with only convective heat losses from the surface.

The poor agreement between the experimental results and curve 3 may be attributed partially to the isothermal pellet assumption used in obtaining this curve. The significance of this assumption is shown in curve 2 which is the sharp interface model using the isothermal pellet assumption but without a radiation heat loss term. It can be seen that neglect of internal temperature gradients (even if these are only 10°C or less) causes a significant reduction in the maximum temperature attained.

To summarize this section, it would appear that simple models based on production of carbon oxides alone cannot give wholly accurate predictions of both the magnitude of the temperature rise and the time at which this would occur. It seems evident that account must be taken of the composition of the coke deposit to obtain a more accurate prediction of the temperature rise developed during coke oxidation from catalysts.

Coke deposits on catalysts generally have a composition in the range $CH_{0.4}$ to $CH_{1.0}$. Massoth (1967) first recognized the importance of this effect. His own results and those of other workers showed that the water produced from the oxidation of hydrogen tended to be produced earlier in the reaction than oxides of carbon from oxidation of the carbon itself. As water vapour production is also exothermic, neglect of this early contribution to the total heat evolution could be the cause of the delayed temperature maxima predicted by the conventional models, especially at lower temperatures. Massoth suggested that, because of the faster reaction of the hydrogen in the coke, the hydrogen reacts at an inner interface of a catalyst particle leaving a dehydrogenated residue of carbon in the outer layers which reacts with oxygen at a lower rate. The difficulty Massoth encountered in attempting to model his experimental results was overcome by Ramachandran *et al.* (1975) who assumed that the carbon reacts in the zone between the surface and the sharp interface where the hydrogen reacts. This assumption is reasonable in view of the much faster rate of hydrogen oxidation observed experimentally, and should be applicable at all temperatures except those for which complete diffusional mass transfer of oxygen occurs. Because of the assumptions made, the model is applicable only for the early stages of oxidation; this however is the period during which the maximum temperature rise occurs, so this restriction is not important when the aim is to predict this maximum temperature rise.

Because of the faster reaction of hydrogen in the coke, two zones develop in the pellet, an inner zone having the composition of the original coke and the outer zone being coke from which the hydrogen has been completely removed (Fig. 9.13). The reaction of hydrogen occurs at the interface, λ, between these zones and is assumed to occur via the sharp interface model. The carbon in the coke is assumed to react via a homogeneous model in the zone R to λ. Thus the model envisages diffusion and reaction at a moving boundary coupled with reaction in the diffusion zone.

An important assumption in the model is that the pellet is isothermal, with all the heat transfer resistance localized in the gas film surrounding the pellet. The pseudo-steady approximation was used for the material balances. The heat balance is written in a transient form, however, for the reasons noted previously.

From Fig. 9.13 the differential equation governing the reaction of carbon

Deactivation of Catalysts

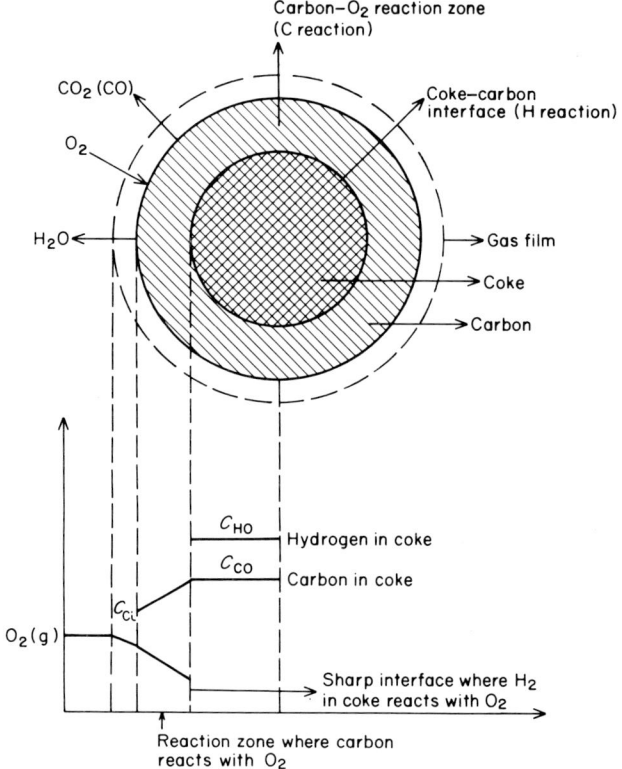

Fig. 9.13 Model for simultaneous carbon and hydrogen reaction. (Ramachandran et al., 1975)

in the coke with diffusing oxygen may be written in the form:

$$\frac{D_e}{r^2}\frac{d}{dr}\left(r^2\frac{dC_o}{dr}\right) = k_c(T)C_cC_o \qquad (R > r > \lambda) \qquad (9.18)$$

where C_o is the oxygen concentration at any radial position r in the pellet, C_c is the concentration of carbon at radial position r, k_c is the second order rate constant for reaction of carbon in the coke with oxygen, T is the temperature of the pellet at any given time.

The hydrogen in the coke is assumed to react with oxygen at a sharp interface, λ. Thus the reaction with hydrogen appears as a boundary condition in this model:

$$D_e\frac{dC_o}{dr} = k_H(T)C_o \qquad (\text{at } r = \lambda) \qquad (9.19)$$

where k_H is the surface reaction rate constant for hydrogen.

9 Regeneration of Deactivated Catalysts

The second boundary condition necessary for the solution of equation (9.18) can be formulated by considering the flux of oxygen through the gas film:

$$D_e \frac{dC_o}{dr} k_g(C_{og} - C_o) \quad \text{(at } r = R\text{)}$$

where C_{og} is the concentration of oxygen in the bulk gas and k_g is the gas film mass transfer coefficient.

Equation (9.18) and the associated boundary conditions can be reduced to dimensionless form by introduction of the following parameters:

$$\psi = \frac{C_o}{C_{og}}; \quad \delta = \frac{r}{R}; \quad \phi_o^2 = \frac{k_c^* C_{co} R^2}{D_e}$$

$$\text{Sh}^* = \frac{k_g R}{D_e}; \quad B^* = \frac{k_H^* R}{D_e}; \quad \lambda_d = \frac{\lambda}{R}$$

$$\theta = \frac{T - T_g}{T_g}; \quad \gamma_1 = \frac{E_1}{RT_g}; \quad \gamma_2 = \frac{E_2}{RT_g}$$

where C_{co} is the initial concentration of carbon, k_c^* and B^* are rate constants based on the bulk gas temperature T_g, E_1 and E_2 are activation energies for reaction of carbon and hydrogen respectively.

Equation (9.18) now becomes:

$$\frac{1}{\delta^2} \frac{d}{d\delta}\left(\delta^2 \frac{d\psi}{d\delta}\right) = \phi_o^2 \frac{C_c}{C_{co}} \psi \exp\left[\gamma_1\left(1 - \frac{1}{1+\theta}\right)\right] \quad \text{(for } 1 > \delta > \gamma_d\text{)} \quad (9.20)$$

with the boundary conditions:

$$\frac{1}{\text{Sh}^*} \frac{d\psi}{d\delta} = 1 - \psi \quad \text{(at } \delta = 1\text{)} \quad (9.21)$$

$$\frac{1}{B} \frac{d\psi}{d\delta} = \psi \quad \text{(at } \delta = \lambda_d\text{)} \quad (9.22)$$

where

$$B = B^* \exp\left[\gamma^2\left(1 - \frac{1}{1+\theta}\right)\right]$$

Equation (9.20) can be solved only numerically, since the carbon concentration C_c varies with the radial position δ. However, in the initial periods of regeneration, the reaction of carbon is not significant and the variation of carbon concentration in the zone λ_d to 1 is not large. Hence the value of C_c can be replaced by an average value $\bar{C}_c$ for the region λ_d to 1. The value of $\bar{C}_c$ is taken as:

$$\bar{C}_c = (C_{co} + C_{ci})/2 \tag{9.23}$$

where C_{ci} is the interfacial concentration of carbon. The value of C_{ci} can be predicted by a simple mass balance as shown later in this chapter. With the above approximation, equation (9.20) becomes:

$$\frac{1}{\delta^2}\frac{d}{d\delta}\left(\delta^2 \frac{d\psi}{d\delta}\right) = \phi^2 \psi \tag{9.24}$$

where

$$\phi^2 = \phi_o^2 \frac{\bar{C}_c}{C_{co}} \exp\left[\gamma_1\left(1 - \frac{1}{1+\theta}\right)\right]$$

The solution of equation (9.24) with the boundary conditions (9.21) and (9.22) gives the concentration profile of oxygen in the pellet:

$$\psi = \frac{\alpha}{\delta} \sinh[\phi(\delta - \lambda_d)] + \frac{\beta}{\delta} \cosh[\phi(\delta - \lambda_d)] \tag{9.25}$$

where α and β are defined as:

$$\frac{Sh^*}{\beta} = \cosh[\phi(1 - \lambda_d)]\left(B + \frac{1}{\lambda_d} - 1 + Sh^*\right)$$

$$+ \sinh[\phi(1 - \lambda_d)]\left(\frac{BSh^*}{\phi} - \frac{B}{\phi} - \frac{1}{\lambda_d \phi} + \frac{Sh^*}{\lambda_d \phi} + \phi\right) \tag{9.26}$$

and

$$\lambda = \frac{\beta}{\phi}\left(B + \frac{1}{\lambda_d}\right) \tag{9.27}$$

The net rate of reaction of oxygen with hydrogen in the coke at any time is given by the diffusional flux of oxygen at λ or:

$$R_{H_2} = 4\pi\lambda^2 D_e \left(\frac{dC_o}{dr}\right)_{r=\lambda} \tag{9.28}$$

It is convenient to express the rate of reaction as the ratio of the actual rate of reaction to a reference rate defined as:

$$R_{ref} = \tfrac{4}{3}\pi R^3 k_c^* C_{co} C_{og} \tag{9.29}$$

Using the value of (dC_o/dr) obtained from differentiating equation (9.25)

9 Regeneration of Deactivated Catalysts

we obtain:

$$\eta_{H_2} = \frac{R_{H_2}}{R_{ref}} = \frac{3}{\phi_o^2} \lambda_d \left(\alpha\phi - \frac{\beta}{\lambda_d} \right) \quad (9.30)$$

The net reaction of oxygen with carbon is:

$$R_c = 4\pi R^2 D_e \left(\frac{dC_o}{dr}\right)_{r=R} - 4\pi\lambda^2 D_e \left(\frac{dC_o}{dr}\right)_{r=\lambda} \quad (9.31)$$

The values of the diffusive fluxes of oxygen at $r = R$ and λ can again be obtained from equation (9.25), and hence we obtain:

$$\eta_c = \frac{R_c}{R_{ref}} = \frac{3}{\phi_o^2} \left\{ \left(\alpha\phi - \frac{\beta}{\lambda_d} \right) [\cosh[\phi(1-\lambda_d)] - 1] \right.$$

$$\left. + (\beta\phi - \lambda)[\sinh[\phi(1-\lambda_d)]] \right\} \quad (9.32)$$

The change in λ_d, the position of "oxygen–hydrogen reaction" interface with time, can be obtained by writing a mass balance for net hydrogen reacted with oxygen:

$$4\pi\lambda^2 D_e \left(\frac{dC_o}{dr}\right)_{r=\lambda} = \frac{1}{2}\frac{d}{dt}(\tfrac{4}{3}\pi\lambda^3 C_{Ho}) \quad (9.33)$$

where C_{Ho} is the initial concentration of hydrogen in the coke. Using equations (9.28) and (9.30), the following equation can be obtained for the change in λ_d with time:

$$\frac{d\lambda_d}{d\tau} = \frac{2}{q_H \lambda_d} \left(\alpha\phi - \frac{\beta}{\lambda_d} \right) \quad (9.34)$$

where τ is the dimensionless time $(= tD/R^2)$ and $q_H = (C_{Ho}/C_{og})$.

The transient heat balance equation, assuming no internal temperature gradients in the pellet, is written as:

$$\tfrac{4}{3}\pi R^3 \rho c_p \frac{d(T - T_g)}{dt} = -4\pi R^2 h(T - T_g) + R_{H_2}\Delta H_2 + R_c \Delta H_c \quad (9.35)$$

where ΔH_2 and ΔH_c are the heats of reaction of 1 mole of oxygen with hydrogen and carbon respectively.

Equation (9.35) can be expressed in dimensionless form as:

$$\frac{d\theta}{d\tau} = -3\text{NuLe}\theta + \phi_o^2 \eta_{H_2}\beta_H + \phi_o^2 \eta_c \beta_c \quad (9.36)$$

where θ is the dimensionless temperature $(T - T_g)/T_g$. (The other dimensionless groups are defined in the notation.) Equations (9.34) and (9.36) can be integrated by any standard method, e.g. the Runge–Kutta method, finite difference, etc. The values of η_{H_2} and η_c required for the solution can be obtained from equations (9.30) and (9.32) respectively.

For the solution of equations (9.30) and (9.32) the interface concentration of carbon is required. This can be obtained by a mass balance for carbon for the zone $R > r > \lambda$. For this purpose we assume that C_c varies linearly in the zone λ to R from a value of C_{ci} at $r = R$ to C_{co} at $r = \lambda$. The assumption would hold for up to 50% burn-off for spherical particles. The carbon concentration profile is thus:

$$C_c = C_{co} + \frac{C_{ci} - C_{co}}{(1 - \lambda_d)} (\delta - \lambda_d) \qquad (9.37)$$

The material balance for carbon is:

$$\tfrac{4}{3}\pi R^3 C_{co} - \tfrac{4}{3}\pi \lambda^3 C_{co} - \int_\lambda^R C_c 4\pi r^2 \, dr = \int_0^1 R_c \, dt \qquad (9.38)$$

Using the value of C_c from equation (9.37) in equation (9.38) and integrating, we obtain:

$$\frac{C_{ci}}{C_{og}} = \frac{C_{co}}{C_{og}} + \frac{(1 - \lambda_d) \int_0^\tau \eta_c \phi_0^2 \, d\tau}{\lambda_d \left(1 - \dfrac{\lambda_d^2}{4}\right) - \dfrac{3}{4}} \qquad (9.39)$$

When conventional models which assumed that the coke was composed entirely of carbon were compared with the experimental results of Shettigar and Hughes (1972b) it was observed that the greatest discrepancy occurred at lower temperatures ($\sim 450°C$). Not only was the maximum temperature rise not predicted accurately but the initial rapid temperature rise was not predicted at all; instead a relatively slow approach to the maximum was predicted as the reaction time increased (Fig. 9.11). At the time the model was first developed no precise kinetic data were available for the hydrogen oxidation, so it was assumed to be first order in oxygen and zero order with respect to hydrogen in the coke. Subsequent experimentation (Parvinian, 1977) confirmed that this was true and also that the ratio of the dimensionless rate constants for the hydrogen and carbon oxidation reaction, $k_H^*/k^* C_{co} R$, could be assumed to be equal to 5.

Predicted temperature profiles for the C/H ratios 1/1 and 1/0.5 are compared with experimental results in Figs 9.14 and 9.15 for initial gas temperatures of 450 and 540°C respectively. The data used for the

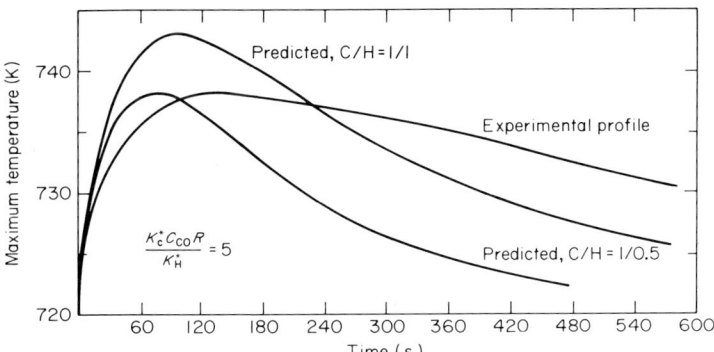

Fig. 9.14 Comparison of predicted temperature profiles at two carbon/hydrogen ratios with experimental results at 723 K; coke level 1%. (Ramachandran *et al.*, 1975)

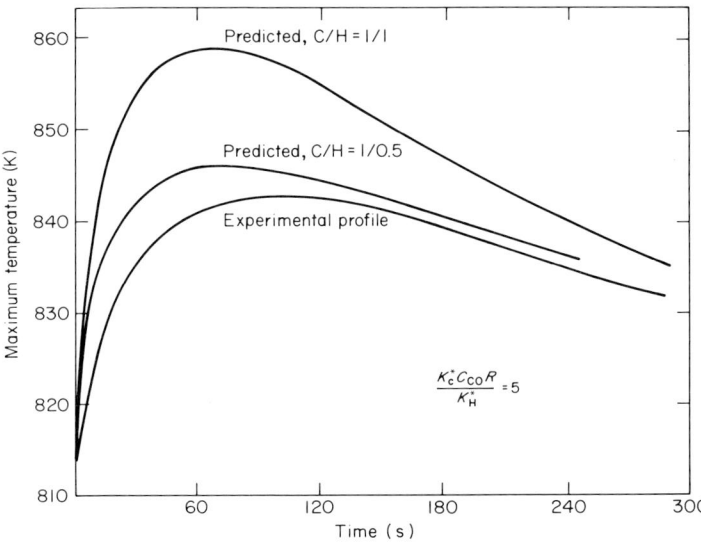

Fig. 9.15 Comparison of predicted temperature profiles at two carbon/hydrogen ratios with experimental results at 813 K; coke level 1%. (Ramachandran *et al.*, 1975)

comparison are given in Table 9.4. It can be seen that the initial temperature maximum occurring early in the reaction at 450°C is now predicted for both carbon/hydrogen ratios. At both bulk gas temperatures the predicted curves agreed more closely with experiment when the carbon/hydrogen ratio is 1/0.5. Since no analyses of coke composition were done in the work of Shettigar, it is not possible to verify whether this was the actual coke

Table 9.4 Physicochemical data used in the computation of the temperature profiles

$C_{og} = 3$ mol m^{-3}	$C_{co} = 780$ mol m^{-3}
$C_{Ho} = 390$ mol m^{-3}	$D_e = 10^{-6}$ m^2 s^{-1}
$\Delta H_2 = 468$ kJ mol^{-1}	$\Delta H_c = 393$ kJ mol^{-1}
Le $= 0.2$	Nu* $= 0.25$
$k_c^* = 1.33 \times 10^7 \exp\left(-\dfrac{18\,900}{T_g}\right)$	
$k_H^*/k_c^* C_{co} R = 3, 5,$ or 7.5	Sh* $= 175$
$\rho = 1.33 \times 10^6$ mol m^{-3}	$c_p = 1.5 \times 10^{-3}$ kJ g^{-1} K^{-1}

composition obtained experimentally. Thus it may be concluded that, although complete correspondence does not occur, this model does give much better agreement with experiment than models that consider coke to be composed of carbon alone, and therefore only models that account for the composition of coke can give reliable predictions of the temperature rise at low regeneration temperatures.

9.5 Regeneration of fixed beds containing coked catalyst

The literature on the regeneration of coked catalysts contains a number of articles on the modelling and prediction of the temperature rise and conversion of coke but there is a paucity of experimental data. There is some information available from commercial practice especially on the regeneration of catalytic reformers used for the processing of low octane straight run naphthas to high octane derivatives. These reactors operate with a platinum on alumina catalyst, so parallel problems of sintering and poisoning also occur during the coking period. However, the regeneration procedure for such catalysts is fairly well established. The procedure is to regenerate with an inert recirculating gas containing 0.5–1.0 mol % of oxygen, and ignition is often obtained at the lowest practical temperature which may be around 370°C. In such operations the coke burn-off is controlled by air injection rate in order to avoid localized burning and to keep the temperature rise across each reactor bed to less than 80°C. On completion of the primary coke burn-off, a proof burn or "soak" may often be required for this particular catalyst. This is a burn-off at a higher temperature and oxygen level to achieve reoxida-

9 Regeneration of Deactivated Catalysts

tion of the platinum surface and to retard the agglomeration of the platinum crystallites.

Since appreciable temperature rises can be obtained in single catalyst pellets it is instructive to consider in general terms the effects that result when a packed bed of catalyst is regenerated. It has frequently been observed that the actual burning of the coke during the regeneration in a fixed bed is limited to a fairly narrow reaction zone which passes through the bed and gives rise to a temperature wave. The regenerative heating of the combustion gas, as it passes over that portion of the bed from which the coke has already been burned, results in a build-up of the temperature profile as this moves through the bed. Thus the temperatures encountered can be considerably in excess of those originating in the single pellets alone. In many cases the design of the reactor for both the main reaction and the regeneration process is determined largely by the regeneration aspect, since this is the more crucial. One important consideration in systems with a sharp moving front is the relative positions of the actual reaction zone and the temperature front. Owing to the properties of the bed itself, including the effective bed thermal conductivity and the various heat transfer processes at the reactor wall, these are not necessarily moving together and this can have important consequences in both operation and design. The majority of workers have limited their terms of reference to adiabatic conditions and to a simple plug flow reactor model with no density gradients. Additionally, many have assumed that heat transfer between the catalyst pellets and the gas is sufficiently great to ensure that the temperature difference between the two phases is negligible. However, as shown in the preceding section, this is not generally true. Another assumption frequently used is that the product gas consists only of carbon dioxide. In some ways this is justifiable since it presents the worst possible case because the heat of reaction assuming carbon dioxide as the sole product is larger than when the contributions due to carbon monoxide and water vapour production are included.

Most of the work published on regeneration of fixed beds has concentrated on the temperature rise encountered. This is understandable in view of the effect of temperature rise on catalyst activity and selectivity. Because the temperature rise is a function of both temperature and position in the bed, three separate temperature maxima may be defined. These are the maximum temperature at a particular position in the bed at any time during the regeneration, the maximum temperature at a particular time along any part of the bed, and finally a "peak" temperature which can be defined as the highest temperature attained during the entire process at any position and any time.

The temperature rise developed in an adiabatic fixed bed during regeneration reaches a maximum in the early stages of the process when only a small part of the bed has been regenerated. This has been attributed by Olson et al. (1968) to the high initial rate of reaction produced when the relatively rich gas contacts the coked catalysts. The rapid premature development of the temperature peak attains an asymptotic maximum value for the rest of the diffusion controlled reaction. This asymptotic maximum value is dependent on both initial oxygen and coke concentrations.

One possibility which must be considered is that of temperature "runaway". The combustion zone or zone of reaction moves along the bed with a velocity determined by the reaction rate and the initial coke and oxygen concentrations. The heat front resulting from the exothermicity of the combustion of the coke moves independently of the combustion zone because of the convective effect of gas flow. When the combustion zone and heat front progress at the same speed the heat of reaction accumulates within the combustion zone and the temperature of this zone will tend towards infinity. Critical values of both initial coke and oxygen concentrations exist at which this phenomenon can occur. Since the coke concentration at which regeneration should begin is usually predetermined, the inlet oxygen concentration becomes the crucial variable.

Attempts have been made to limit the maximum temperature rise by various procedures. The use of steam in the regenerator to limit the temperature rise seems to be of doubtful value, while the dilution of the catalyst bed using metals may have unusual and opposite effects as discussed later. Since both CO and CO_2 are the primary products of carbon oxidation, efforts to minimize secondary oxidation of CO to CO_2 are essential to control temperature rises. A similar situation which occurs in fluidized bed regenerators has already been mentioned.

For a fixed coke level the asymptotic temperature rise is dependent on the inlet oxygen concentration. Also, the rapid temperature rise observed is an initial phenomenon. It has been suggested therefore that regeneration should be carried out at low inlet concentrations of oxygen initially, subsequently increasing the oxygen level at later stages of the regeneration process. This procedure is adopted in many industrial processes.

The first analysis of the temperature "runaway" problem was made by Thompson (1937) who considered the regeneration of a fixed bed catalytic cracker. This was followed by Van Deemter (1953) and by Johnson et al. (1962). In all these cases certain simplifying assumptions were made in order to obtain analytical solutions. This has the disadvantage of neglecting some realism but does enable trends and limits to be explored.

Thompson (1937) assumed no heat conduction in either the axial or radial directions, and that the rate constant for carbon oxidation was infinite, i.e. a sharp burning front was formed limited only by diffusion of oxygen. From this basis Thompson showed that, for an infinitely long bed, the maximum

9 Regeneration of Deactivated Catalysts

temperature rise could be expressed by

$$\Delta T_{max} = \frac{\Delta H y_o / c_s}{\theta_B / \theta_H - 1} \tag{9.40}$$

where ΔH is the heat of combustion of carbon, y is the initial weight % of carbon on the catalyst, and c_s the catalyst specific heat. The time θ_B required to burn the carbon from the catalyst bed if the oxygen is completely consumed is given by:

$$\theta_B = \frac{\rho_s L y_o M}{12 u \rho_g x_o} \tag{9.41}$$

and θ_H, the transit time for the heat front in the bed, by:

$$\theta_H = \frac{\rho_s L c_s}{u \rho_g c_g} \tag{9.42}$$

In equations (9.41) and (9.42), x_o is the inlet mole fraction of oxygen, L the bed length, M the gas molecular weight, and ρ_g and ρ_s the density of gas and solid phase respectively.

Thompson showed that three possibilities existed for a fixed bed:

If $\theta_B > \theta_H$ $T_{max} = T_{g,o} + \Delta T_{max}$

If $\theta_B < \theta_H$ $T_{max} = T_{g,o} - \Delta T_{max}$

If $\theta_B = \theta_H$ $T_{max} \to \infty$

The latter equality signifies instability, the temperature in a narrow zone increasing continuously by the combustion reaction occurring in the same zone. An important consequence of this analysis is that instability can still occur even when the amount of carbon on the catalyst is low, provided that the inlet oxygen concentration is also low. However, insertion of finite kinetics and a finite bed length limit the maximum temperature predicted.

Van Deemter (1953) in his analysis assumed a zero order reaction and a constant reaction rate. Surprisingly, in view of this gross oversimplification, results were obtained which depicted the qualitative behaviour of the regeneration process reasonably well.

Van Deemter's analysis was based on a homogeneous model with separate balances for oxygen and coke together with an overall heat balance.

The oxygen balance in the reaction is:

$$\varepsilon \frac{\partial C_{O_2}}{\partial t} + u \frac{\partial C_{O_2}}{\partial l} = -r_{O_2} = -R^* \tag{9.43}$$

and for the coke:

$$r_c = -r_{O_2} = -R^* \text{ (per mole of } O_2\text{)} \tag{9.44}$$

where C_{O_2} is the oxygen concentration, u the superficial gas velocity, l the bed length, and ε the porosity.

The heat balance is:

$$[(1-\varepsilon)\rho_s c_s + \varepsilon\rho_g c_g]\frac{\partial T}{\partial t} + \rho_g c_g u \frac{\partial T}{\partial l} = R^*(\Delta H) \tag{9.45}$$

where c_g, c_s and ρ_g, ρ_s are the specific heat and density of gas and solid respectively.

Boundary conditions for the initial stages are:

$$\text{At } l = 0 \qquad C_{O_2} = C_{O_2}^o$$
$$T = T_o \tag{9.46}$$
$$\text{At } t = 0 \ (l > 0) \quad C_{O_2} = 0$$
$$T = 0$$

Equation (9.45) may be solved using the method of characteristics to give:

$$T = T_o + R^*(\Delta H)l/\rho_g c_g u \quad \text{(for } 0 < l < \alpha u t\text{)} \tag{9.47}$$

where

$$\alpha = \frac{\rho_g c_g}{(1-\varepsilon)\rho_s c_s + \rho_g c_g}$$

As equation (9.47) shows, the temperature profile is linear. The solution under these initial conditions is confined to the inlet region of the bed ($l < \alpha u t$), and the heat produced in this region is transported through the bed with a velocity αu. For $l > ut$ the temperature must correspond to the boundary condition $T = 0$. For the same initial operation the solution for the oxygen balance, equation (9.43), is:

$$C_{O_2} = C_{O_2}^o - R^* l/u \quad \text{(for } 0 < l < \alpha u t\text{)} \tag{9.48}$$

The coke distribution follows from the definition of R^* in equation (9.44):

$$C_o = C_o^o - R^*(t - l/\alpha u) \tag{9.49}$$

Thus these material balances also yield linear profiles.

How the system behaves depends entirely on the inlet oxygen concentration, $C_{O_2}^o$. The time for complete coke removal is given by equation (9.49):

$$t_1 = C_o^o/R^* \tag{9.50}$$

9 Regeneration of Deactivated Catalysts

Oxygen is also completely removed [equation (9.48)] when:

$$l_0 = C_{O_2}^o u / R^* \qquad (9.51)$$

corresponding to a time:

$$t_o = l_o/\alpha u = C_{O_2}^o/\alpha R^* \qquad (9.52)$$

When $t_o > t_1$ the coke at the bed entrance will have reacted completely, while oxygen is still present. When $t_1 > t_o$, oxygen becomes the limiting reactant and the combustion zone becomes stabilized until the coke disappears.

The limiting cases were well reproduced by Van Deemter, within the limitations of his model, and thus support the earlier work of Thompson (1937) and stress the importance of the movement of combustion and heat fronts in the bed. The analysis was subsequently extended (Van Deemter, 1954) for low oxygen concentrations. Unfortunately the model predicts an intermittent type of movement of the reaction zone which is of course quite unrealistic; nevertheless the model is of use in explaining and interpreting the limiting operating conditions.

Johnson et al. (1962) extended the analysis of Van Deemter and removed the restrictions of zero order kinetics. Instead the rate controlling mechanism was assumed to be the diffusion of oxygen within the pellet, which implies that the rate is proportional to the partial pressure of oxygen. This has the advantage that the temperature dependence of the diffusion controlled reaction is expected to be low, so that a constant value for the rate coefficient may be assumed without serious error, thus facilitating the solution of the balances. An approximate indication of the increasing diffusion distance is given by the dimensionless carbon concentration ratio C_c/C_{co}, so the rate expression can be written as:

$$r = kPy \frac{C_c}{C_{co}} \qquad (9.53)$$

where C_{co} is the initial carbon content and y the mole fraction of oxygen in a total pressure P of gas.

The appropriate balances for an adiabatic plug flow reactor are:

$$\varepsilon \frac{\rho_g}{M_g} \frac{\partial y}{\partial t} + \frac{G}{M_g} \frac{\partial y}{\partial z} = -\frac{akP\rho_s}{M_c} y \frac{C_c}{C_{co}} \qquad (9.54)$$

with $y = y_o$ at $z = 0$.

In equation (9.54), a is a stoichiometric factor which allows for the possibility of CO production as well as CO_2; it has a value unit for CO_2 production alone. M_g and M_c are the molecular weights of gas and carbon respectively.

Deactivation of Catalysts

Carbon balance:

$$\frac{\partial C}{\partial t} = -kPy\frac{C_c}{C_{co}} \tag{9.55}$$

with $C_c = C_{co}$ at $t - z(\varepsilon\rho g/G) \leq 0$.

Heat balance:

$$[(1-\varepsilon)\rho'_s c_s + \varepsilon\rho_g c_g]\frac{\partial T}{\partial t} + c_g G\frac{\partial T}{\partial z} = \rho_s(-\Delta H)kPy\frac{C_c}{C_{co}} \tag{9.56}$$

The term $(1-\varepsilon)\rho'_s c_s$ is usually much greater than $\varepsilon\rho_g c_g$, so the latter may be neglected, and therefore $(1-\varepsilon)\rho'_s$ may be written as ρ_s, the bed bulk density.

Equations (9.54)–(9.56) may be written in dimensionless form as:

$$\frac{\partial y}{\partial \xi} = -ay\frac{C_c}{C_{co}} \tag{9.57}$$

$$\frac{y_o}{C_o}\frac{\partial C_c}{\partial \tau} = -y\frac{C_c}{C_{co}} \tag{9.58}$$

$$M\frac{\partial T^*}{\partial \tau} + 1 + \frac{\partial T^*}{\partial \xi} = \frac{yC_c}{y_o C_{co}} \tag{9.59}$$

where

$$y = y_o \quad \text{for } \xi = 0 \text{ for all } \xi$$
$$C = C_o \quad \text{at } \tau = 0 \text{ for all } \xi$$
$$T^* = 0 \quad \text{at } \xi = 0$$

while

$$H = \frac{M_g}{M_c}\frac{c_g}{c_s}\frac{C_{co}}{y_o}; \quad M = 1 - \frac{c_g \varepsilon \rho_g}{c_s \rho_s}$$

$$T^* = T\frac{c_s}{(-\Delta H)C_{co}}; \quad \tau = \left(\frac{y_o}{C_{co}}kP\right)\left(t - \frac{z\varepsilon\rho_g}{G}\right)$$

$$\xi = z\frac{\rho_s}{G}\frac{M_g}{M_c}kP$$

Since the reaction rate is independent of temperature, the mass balances may be solved separately to give:

$$\frac{C_o}{C_{co}} = \frac{1}{1 + e^{-a\xi}(e^\tau - 1)} \tag{9.60}$$

$$\frac{y}{y_o} = \frac{1}{1 + e^{-\tau}(e^{a\xi} - 1)} \tag{9.61}$$

These relations give continuous concentration profiles for both carbon and oxygen within the bed, and Johnson et al. (1962) found that these moved as fronts towards the exit of the bed with increasing time. After a certain time these concentration profiles plus the corresponding temperature profiles developed a fixed shape which moved through the bed at a rate determined by the rate of oxygen supply. In order to interpret the heat balance the second terms within the parentheses of equations (9.60) and (9.61) are neglected. This is equivalent to assuming that a fixed profile is developed at the beginning of the reaction. Equations (9.60) and (9.61) then reduce to:

$$\frac{C_c}{C_{co}} = \frac{1}{1 + \exp(-a\xi + \tau)} \tag{9.62}$$

$$\frac{y}{y_o} = \frac{1}{1 + \exp(a\xi - \tau)} \tag{9.63}$$

and their product, which is a measure of the rate of heat release:

$$\frac{yC_c}{y_o C_{co}} = \frac{1}{2 + \exp(a\xi - \tau) + \exp(-a\xi + \tau)} \tag{9.64}$$

Figure 9.16 compares the exact solution with the above approximate solutions for the heat release (product of oxygen and carbon concentrations in the bed) as a function of position. It can be seen that the two solutions are virtually identical with increasing distance from the inlet of the bed.

Introduction of equation (9.64) into the heat balance enables this to be solved analytically using the method of characteristics, to yield finally:

$$T^* = \frac{1}{2(aH - M)} \left\{ \tanh\left(\frac{a\xi - \tau}{2}\right) - \tanh\left[\frac{\xi - (H/M)\tau}{2H/M}\right] \right\} \tag{9.65}$$

From this, temperature profiles were predicted as shown in Fig. 9.17 for different times. The passage of the temperature peak down the bed is clearly indicated.

In addition Johnson et al. (1962) also experimentally measured the temperature profiles during coke regeneration. The catalyst was pre-coked by immersing in sugar solutions followed by thermal treatment. Adiabatic conditions were not achieved under any conditions, and in some experiments coolant was circulated under isothermal conditions to achieve significant wall heat transfer. No details of catalyst particle size were given. The

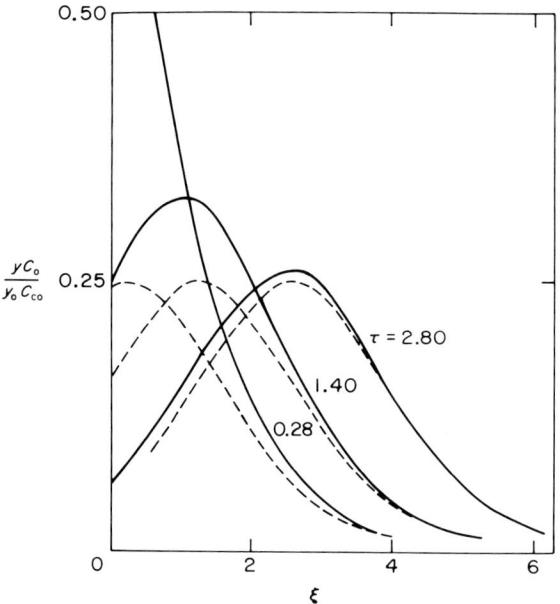

Fig. 9.16 Variation in the product of oxygen and carbon concentration down the bed: (———) exact solution, (– – – –) approximate solution. (Johnson *et al.*, 1962) $G = 1.36$ kg s^{-1} m^{-2}, $C_{CO} = 0.02$ kg (kg catalyst)$^{-1}$, $y_o = 0.02$, $\rho_s = 960$ kg m^{-3}, $c_g = 1.05$ kJ kg^{-1} K^{-1}, $c_s = 1.21$ kJ kg^{-1} K^{-1}, $kP = 0.0078$ s^{-1}.

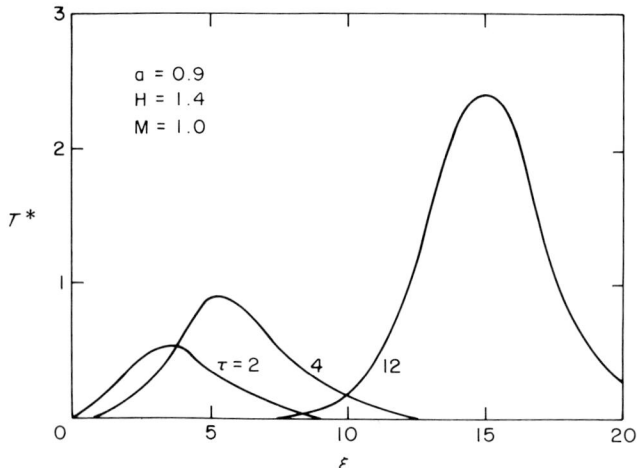

Fig. 9.17 Reduced temperature as a function of position at various times. (Johnson *et al.*, 1962)

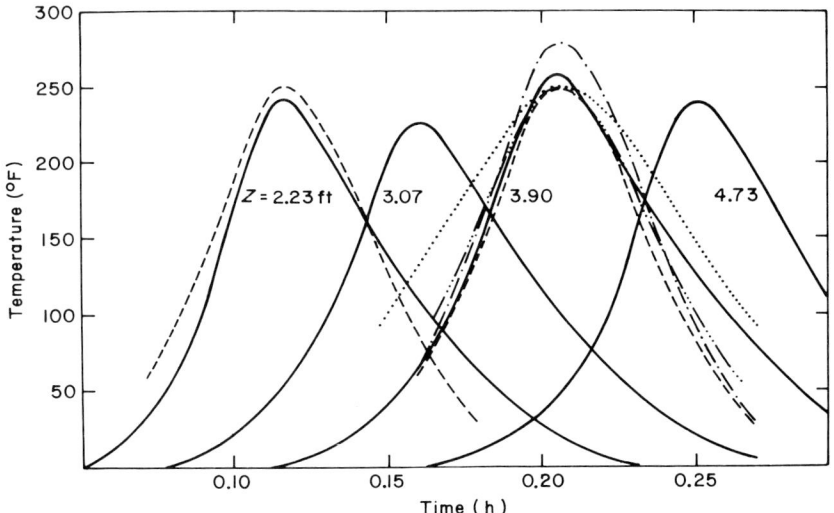

Fig. 9.18 Comparison of experimental (full curves) and calculated (dashed and dotted curves) temperature profiles with isothermal cooling. (Johnson *et al.*, 1962)

experimental temperature profiles obtained as shown in Fig. 9.18 and the definite peaks and their progression along the bed are compared with theoretical predictions. Agreement is seen to be very good, particularly with the peak maximum temperature which was about 130°C above the inlet temperature level.

In contrast to the previous analytical solutions, Schulman (1963) used a numerical solution to solve the appropriate balances. Because of this he was able to remove the restrictions consequent upon the adoption of analytical solutions. Thus the temperature coefficient of the rate constant was now included, as was wall heat transfer, but nevertheless CO_2 was assumed to be the only product. The partial differential equations for temperature, oxygen, and carbon then become:

$$\frac{\partial T}{\partial t} + \frac{u\rho_g c_g}{\rho_s c_s}\frac{\partial T}{\partial z} = (-\Delta H)kPxy - \frac{4h}{d}\frac{(T - T_c)}{\rho_s c_s} \qquad (9.66)$$

$$\frac{\partial y}{\partial z} = -\frac{M\rho_s}{12u\rho_g}kPxy \qquad (9.67)$$

$$\frac{\partial x}{\partial t} = kPxy \qquad (9.68)$$

where y is the mole fraction of oxygen in the gas and x the mole fraction of carbon on the catalyst. Appropriate boundary conditions are:

$$At\ t = 0 \quad T = T_{s,o} \text{ for all } z$$
$$x = x_o \text{ for all } z$$
$$At\ z = 0 \quad T = T_{g,o} \text{ for all } t$$
$$y = y_o \text{ for all } t$$

In the above, $T_{s,o}$ refers to solid temperatures at the inlet and $T_{g,o}$ to the gas inlet temperatures.

Numerical results obtained for adiabatic operation of a catalytic reformer are shown in Fig. 9.19, for the computation parameter values given in Table 9.5.

Table 9.5 Conditions for computations shown in Figs 9.19 and 9.20.

$T_{s,o} = 527°C$	$c_g = 0.06$ kJ kg^{-1} K^{-1}
$T_{g,o} = 527°C$	$c_s = 0.06$ kJ kg^{-1} K^{-1}
$y_o = 0.01$	$u = 0.23$ m s^{-1}
$x_o = 0.066$	$P = 5$ bar
$\rho_s = 0.88 \times 10^{-3}$ kg m^{-3}	$M = 30$
$\rho_g = 3.4 \times 10^{-6}$ kg m^{-3}	$k_o = 0.103$
$L = 2.3$ m	

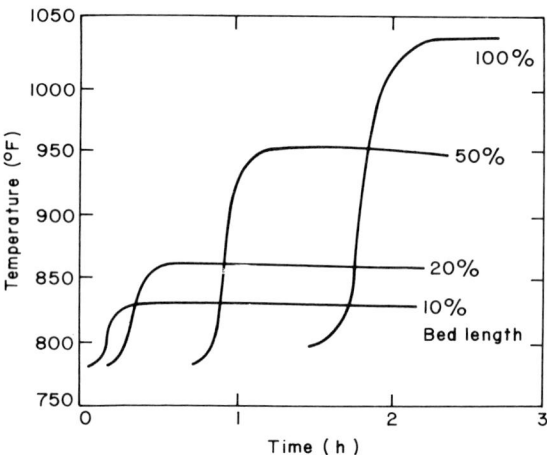

Fig. 9.19 Adiabatic regeneration; development of sharp temperature fronts. Variables have the values listed in Table 9.5. (B. L. Schulman, 1963)

9 Regeneration of Deactivated Catalysts

Sharp temperature fronts develop which move down the bed, and at a given bed position the temperatures reach a maximum and slowly drop as the carbon burning rate drops. The maximum temperature of the front continually rises as the front moves down the bed, since heat generated by the reaction adds continuously to the temperature of the gas entering the reaction zone. The maximum temperature rise was found to be 144°C.

The numerical calculations of Schulman indicated that the temperature front "breaks through" the end of the bed at a time θ_H of about 1.8 h for the case illustrated. The amount of time required to burn all the carbon, assuming 100% utilization of the oxygen, is about 30 h, so at temperature breakthrough the regeneration is only 6% complete. This shows the importance of calculating the time required for complete regeneration independently of the time required to obtain temperature breakthrough.

Schulman also developed a generalized correlation to predict the peak temperatures. A parameter ψ, defined as the ratio of peak temperature rise calculated to the maximum theoretical adiabatic temperature rise, is written:

$$\psi = \frac{T_{\text{peak}} - T_{g,o}}{T_{\max} - T_{g,o}} \quad (\text{for } \theta_B > \theta_H) \tag{9.69}$$

and this, coupled with the dimensionless reaction group γ, where:

$$\gamma = k_o y_o P \theta_B (z/L) \tag{9.70}$$

enables this peak temperature to be estimated from a plot of ψ against γ. This was found to give reasonable estimates of the peak temperature when compared with results from small laboratory reactors which were approximately adiabatic.

When these predictions were compared with results for a commercial catalytic reformer the extent of heat losses was found to be very significant. Equation (9.66) has a wall heat transfer term included to allow for non-adiabatic operation, and a comparison of adiabatic and non-adiabatic (both measured and calculated) is given in Fig. 9.20. The agreement between predicted and observed temperatures is reasonably satisfactory when allowance is made for wall heat transfer using an acceptable value for the wall heat transfer coefficient of 0.68 kw m^{-2} K^{-1}.

In contrast to the models described above, Olson *et al.* (1968) developed a model for regeneration in which separate balances were written for solid and gas phases. The Weisz and Goodwin (1966) shell progressive model was used with a parabolic intraparticle coke concentration profile. The most important conclusion to emerge from this analysis is that large temperature transients can occur during the initial stages of bed regeneration. These are caused by the initial high rate of reaction and rate of heat generation arising when a rich oxygen gas encounters the outer carbon-rich surface of an

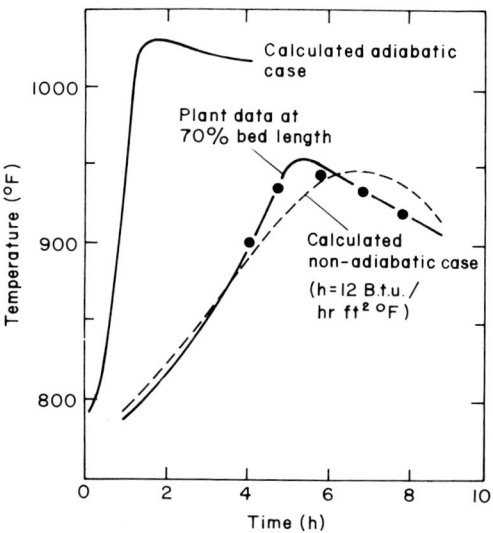

Fig. 9.20 Heat losses included for accurate prediction of temperatures. Conditions are the same as for Fig. 9.19. (B. L. Schulman, 1963)

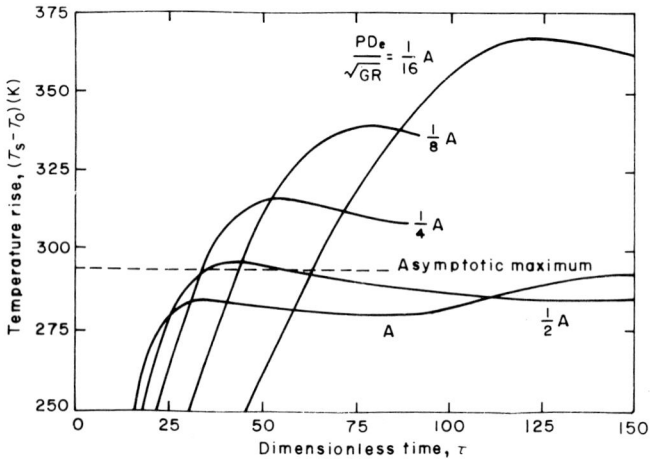

Fig. 9.21 Initial maximum temperature transients; diffusion mass transfer as parameter. (Olson *et al.*, 1968) $A = 1.35 \times 10^{-5}$ bar m$^{5/2}$ (kg s K)$^{-1/2}$

unregenerated catalyst. The initial temperature transients are shown in Fig. 9.21 and it can be seen that, as the parameter A is decreased, very large temperature transients, well in excess of the asymptotic predicted temperature, develop. The parameter A is expressed as $16PD_e/(GR)^{\frac{1}{2}}$, where P is the total pressure, D_e the effective diffusivity of oxygen within the pellet, R the pellet radius, and G the mass flow rate based on the void volume of the bed. Once the combustion zone is fully developed, the maximum temperature in the bed approaches a constant asymptotic level which is independent of $PD_e/(GR)^{\frac{1}{2}}$. This would normally occur after the first few centimetres of bed had been regenerated. However, at sufficiently low values of $PD_e/(D_e)^{\frac{1}{2}}$ the transient maximum temperature exceeds the maximum. From the definition of this parameter (which is essentially the ratio of internal to external mass transfer) it can be seen that it decreases with increased flow. Thus a decrease in flow rate would cause an increased initial maximum temperature. This is because the initial rate on the surface of the coked particles is proportional to the mass transfer coefficient which varies as $G^{\frac{1}{2}}$. Even though in this case the temperature peak moves faster out of the combustion zone, the initial rate is sufficiently great to cause a higher temperature to be attained before significant heat dissipation occurs. The values used to obtain the curves shown in Fig. 9.21 are given in Table 9.6.

Table 9.6 Parameter values used by Olson *et al.* (1968) for Fig. 9.21.

Total pressure	3 bar
Inlet temperature	300°C
Mole fraction of O_2	0.02
Coke concentration	$0.06[0.067 + (r/R)^2]$
Solid density	1.1×10^{-3} kg m^{-3}
Pellet radius	1.5 mm
Gas flow rate	9.57×10^{-4} kg m^{-2} s^{-1}
D_e	$5.5 \times 10^{-8} \, T^{\frac{1}{2}}$ m^2 s^{-1}
M	30

Olson *et al.* also compared their results with the model of Johnson *et al.* (1962). Surprisingly, although the model of Johnson *et al.* was a homogeneous one which does not allow for different temperatures between gas and solid, excellent agreement was obtained. This indicates that the temperature difference between the two phases is probably small compared with the peak temperatures attained in the burning front.

With increasing use of zeolite type catalysts, which are more sensitive to thermal damage than silica–alumina catalysts, the importance of operating at a lower temperature for regeneration has increased. Ozawa (1969) has simulated this region where inlet temperatures are around 420°C. At these temperatures the burning rate is controlled by the intrinsic chemical kinetics with a much higher activation energy than the diffusion controlled regime. This causes a significant difference in the temperature and concentration profiles, and in particular a minimum in the coke concentration profile was shown to occur.

More recently Sampath *et al.* (1975b) have developed a model based on the particle pellet concept developed by Calvelo and Smith (1970) and by Szekely and Evans (1971). Coke and oxygen levels and the effect of particle size were all considered, as was the composition of the product gases. An interesting feature was an analysis of the effect of perturbations on the overall regeneration process. In actual practice perturbations of the oxygen level in the form of step concentration rises are often employed to reduce the time required for regeneration after the first initial low oxygen concentration period has passed. Details of the perturbations are listed in Table 9.7, and the corresponding maximum temperatures developed are plotted in Fig. 9.22. The

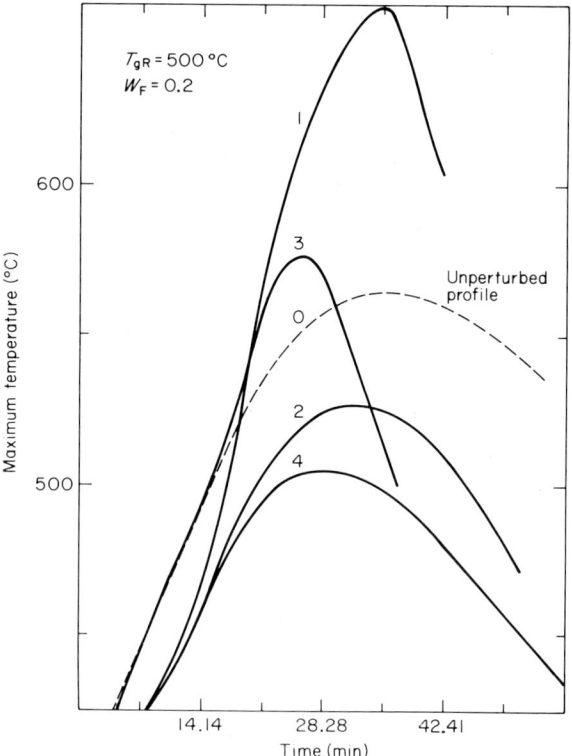

Fig. 9.22 Effect of perturbing reactor inlet conditions for the particle pellet model. (Sampath *et al.*, 1975d)

Table 9.7 Details of perturbation of inlet concentration and temperature

	Dimensionless oxygen concentration of the inlet gas	Dimensionless temperature of inlet gas	Time of reaction at which the perturbation is carried out	Dimensionless oxygen concentration of the inlet gas	Dimensionless temperature of inlet gas	Percentage improvement in conversion over the unperturbed reaction	Peak temperature
	(Initial)	(Initial)	(min)	(Final)	(Final)		(°C)
0.	1.0	1.0	—	1.0	1.0	—	582
1.	0.7	1.0	11.3	1.7	0.95	+48.2	625
2.	0.9	1.0	11.3	1.45	0.95	− 0.5	564
3.	0.95	1.0	11.3	1.35	0.9	+ 8.1	588
4.	0.9	1.0	11.3	1.3	0.95	−11.2	552.5

dashed profile refers to the case where the reactor was operated without any perturbation. Conversion of coke into reaction products was compared over the first 40 min of reaction with all step charges taking place at a time of 11.3 min. Taking each case individually we see that for case 1, where the oxygen concentration at the inlet was increased by 70%, although there was a 50% increase in conversion, the peak temperature is now much higher. Similarly in case 3 the peak temperature is greater than for the unperturbed system although only marginally so. This is expected as the step increase in concentration is now 33%; the percentage increase in conversion is 8% greater than for the unperturbed case. Both case 2 and case 4 shows a reduced temperature profile compared with the unperturbed case which is, however, achieved at the expense of conversion.

Sampath et al. (1975c) have also compared the predictions of the various models on regeneration behaviour. The three models considered were those based on the individual pellet models, namely, the sharp interface retracting core model, the finite thickness moving reaction zone model, and the particle pellet model. All models were developed by writing separate balances for the solid and gas phases. Wide variations in the predicted maximum temperature profiles were observed. As might be expected, the sharp interface retracting core model predicted the highest temperature rises and would find greatest applicability under operating conditions that favour dominant diffusion control of the reaction. The finite thickness moving reaction zone model may be useful for reaction conditions in which both the intrinsic kinetics of the reaction as well as diffusion controls the overall rate. For the parameter values considered, the temperature profiles generated using the particle pellet model lie between those generated by the other two models.

An analysis of non-isothermal systems, in which heat losses could be effected by either radiation or through the reactor wall, was also carried out by Sampath et al. (1975d). It was observed that heat losses through the reactor walls were much more important than the interphase radiation losses between the bulk gas and the solid in the reactor.

The effect of a perturbation in reactor inlet temperatures was also considered. The interesting case was when a step decrease in gas inlet temperature was simulated. This could occur in practice, for example, if a breakdown in the gas preheater occurred. The results obtained for various failure times are illustrated in Fig. 9.23. The sharp interface retracting core model was used for these computations and the dashed line represents the normal (unperturbed) temperature profile. Curves 1, 2, and 3 refer to preheater failure at 3.5, 9, and 23 min respectively from the beginning of the regeneration. In all of these a sharp rise in temperature in the reactor was observed to occur immediately after the negative temperature step. The magnitude of the temperature rise was dependent on the original temperature

9 Regeneration of Deactivated Catalysts

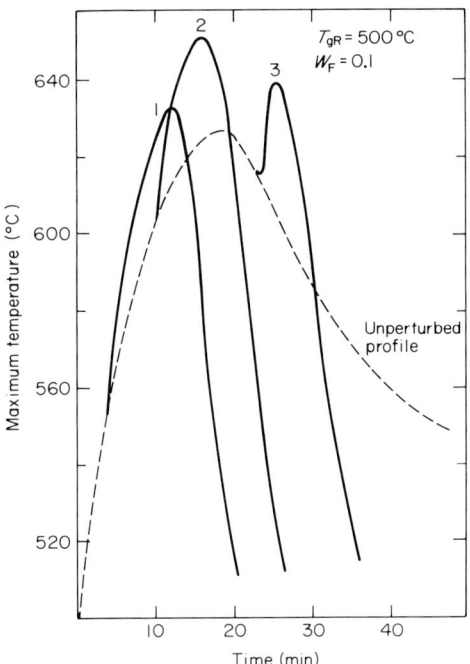

Fig. 9.23 Effect of air preheater failure for the particle pellet model. (Sampath *et al.*, 1975d)

history of the reactor. The rise observed is larger if the temperature is already high. The drop in temperature after this rise is also very steep. This surprising temperature rise, which occurred even though the inlet gas enters the reactor at room temperature after a simulated preheater failure, may be explained as follows. When cool inlet gas enters the reactor it quenches almost all the reaction in this inlet region. However, its passage through this section of the reactor raises its temperature considerably because of the residual heat content of the bed. Thus a hot gas, whose oxygen content is more or less undiminished because of lack of reaction at the bed entrance, meets the unconsumed coke further along the bed. This results in a very rapid reaction developing with a consequent jump in reactor temperature. The cool incoming gas, however, rapidly absorbs the residual heat in the bed and the fast reaction nearer the bed exit is rapidly quenched with a sharp drop in temperature at all points within the reactor. Thus care must be taken in emergency shut-down procedures since, if the gas preheater is allowed to cool rapidly before the reactor is cooled, appreciable temperature rises could occur which could deactivate the catalyst.

References

Ausman, J. M. and Watson, C. C. (1962). *Chem. Eng. Sci.* **17**, 323.
Beveridge, G. S. G. and Goldie, P. J. (1975). *Chem. Eng. Sci.* **30**, 1391.
Bischoff, K. B. (1963). *Chem. Eng. Sci.* **18**, 711.
Bondi, A., Miller, R. S. and Schlaffer, W. G. (1962). *Ind. Eng. Chem.* (*Proc. Des. Devel.*) **1**, 196.
Bowen, J. H. and Cheng, C. Y. (1969). *Chem. Eng. Sci.* **24**, 1829.
Calvelo, A. and Smith, J. M. (1970). *Chemica.*
Dart, J. C., Savage, R. T. and Kilbride, C. G. (1949). *Chem. Eng. Progr.* **45**, 102.
Dobychin, D. B. and Klibanova, Z. M. (1959). *J. Phys. Chem.* (*U.S.S.R.*) **33**, 1023.
Haldeman, R. G. and Botty, M. C. (1959). *J. Phys. Chem.* **63**, 489.
Hughes, R. and Shettigar, U. R. (1971). *J. Appl. Chem.* **21**, 35.
Ishida, M. and Wen, C. Y. (1968). *A.I.Ch.E.J.* **14**, 311.
Johnson, B. M., Froment, G. F. and Watson, C. C. (1962). *Chem. Eng. Sci.* **17**, 835.
Johnson, M. F. L. and Mayland, H. C. (1955). *Ind. Eng. Chem.* **47**, 127.
Luss, D. and Amundson, N. R. (1969). *A.I.Ch.E.J.* **15**, 194.
Massoth, F. E. (1967). *Ind. Eng. Chem.* (*Proc. Des. Devel.*) **6**, 200.
Mickley, H. S., Nestor, J. W. and Gould, L. A. (1965). *Can. J. Chem. Eng.* **43**, 61.
Murphy, J. R. and Soudek, M. S. (1977). *Oil Gas J.* **75**(3), 70.
Olson, K. E., Luss, D. and Amundson, N. R. (1968). *Ind. Eng. Chem.* (*Proc. Des. Devel.*) **7**, 96.
Ozawa, Y. (1969). *Ind. Eng. Chem.* (*Proc. Des. Devel.*) **8**, 378.
Parvinian, M. (1977). Ph.D. Thesis, Salford University.
Ramachandran, P. A., Rashid, M. H. and Hughes, R. (1975). *Chem. Eng. Sci.* **30**, 1391.
Richardson, J. T. (1972). *Ind. Eng. Chem.* (*Proc. Des. Devel.*) **11**, 5.
Rostrup-Nielsen, J. R. (1968). *J. Catal.* **11**, 220.
Rostrup-Nielsen, J. R. (1971). *J. Catal.* **21**, 171.
Sampath, B. S., Ramachandran, P. A. and Hughes, R. (1975a). *Chem. Eng. Sci.* **30**, 125.
Sampath, B. S., Ramachandran, P. A. and Hughes, R. (1975b). *Chem. Eng. Sci.* **30**, 135.
Sampath, B. S., Ramachandran, P. A. and Hughes, R. (1975c). *Trans. Inst. Chem. Eng.* **53**, 234.
Sampath, B. S., Ramachandran, P. A. and Hughes, R. (1975d). *Can. J. Chem. Eng.* **53**, 184.
Schulman, B. L. (1963). *Ind. Eng. Chem.* **55**, No. 12, 44.
Shettigar, U. R. and Hughes, R. (1972a). *Chem. Eng. J.* **3**, 93.
Shettigar, U. R. and Hughes, R. (1972b). *Chem. Eng. J.* **4**, 208.
Szekely, J. and Evans, J. W. (1971). *Chem. Eng. Sci.* **26**, 1901.
Thompson, W. I. (1937). *Calculation of Flame Front Temperatures in Catalytic Cracking Units During Regeneration.* Memo, Aug. 17th.
Van Deemter, J. J. (1953). *Ind. Eng. Chem.* **45**, 1227.
Van Deemter, J. J. (1954). *Ind. Eng. Chem.* **46**, 2300.
Venuto, P. B. and Habib, E. T., Jr. (1979). *Fluid Catalytic Cracking with Zeolite Catalysts.* Marcel Dekker, New York.
Walker, P. L., Rusinko, J., Jr. and Austin, L. J. (1959). *Adv. Catal.* 133.
Weisz, P. B. and Goodwin, R. D. (1963). *J. Catal.* **2**, 397.
Weisz, P. B. and Goodwin, R. D. (1966). *J. Catal.* **6**, 227.
Yagi, S. and Kunii, D. (1961). *J. Chem. Soc. Japan* (*Ind. Chem. Sect.*) **56**, 131.

Index

Activity decay
  correlations for, 49, 50, 51, 147
  effect of Thiele modulus, 152
  expressions for conversion in ageing reactors, 196, 197
  poisoned fixed bed reactors, 152, 153
  shell model, 153
  time dependence in poisoning, 95
  time on stream theory, 116–118
Adsorption, see also Langmuir–Hinshelwood mechanisms
  adsorption constants, single pellet coking, 138, 139
  adsorption constants, packed bed coking, 179–183
  fixed beds, poison distribution, 147
  quinoline, 10, 11
  selective for metal catalyst areas, 57
Alkali, incorporation in catalysts, 10, 11, 111
Alumina
  acid strength, 10
  alumina/boria catalyst, 114–115
  poisoning, 10
  sintering, 3, 24
Ammonia synthesis catalyst, reversible poisoning, 84–85
Atom capture by crystallites, 71
Atomic migration model for sintering, 69

Bed plugging, 21, 140
Bidisperse pore structure, 29, 30
Bifunctional catalysts
  effect of poisons on selectivity, 154
  platinum reforming, 56
  rhodium–silica, mercury poisoning of, 13

Carbon
  burning rate on regeneration, 216
  gas phase, 13
Carbon monoxide oxidation in fixed bed, poisoned by H_2S, 150
Catalytic rich-gas process, 155

Coalescence of metal crystallites, see Particle migration
Coke
  description of deposit, 214
  empirical formula for deposit, 214, 235
  hydrogen constituent, 217, 235
  hydrogen burn-off, 235
Coke formation
  analysis of, isothermal, 119
  deactivation by, theory of, 116
  empirical correlations, 14
  experimental results, 113–116
  modelling in single catalyst pellets, 119
  pore blocking, 18, 19, 112
  precursors, 13, 111, 112
  processes, 112
  profiles, catalyst pellet, 114, 115
Coking
  activity, concentration and temperature profiles in pellets, 121, 125, 126, 132, 136
  acetylene dehydrogenation, 111
  catalytic reforming, 111
  dehydrogenation of n-butyl alcohol, 114–115
  disproportionation of toluene, 114–115
  experimental determination in fixed beds, 113, 160
  experimental determination in catalyst pellets, 114
  modelling in catalyst pellets, 119–140
  modelling in fixed beds using cell model, 163–167
  modelling in isothermal fixed bed reactors, 158–162
  modelling in non-isothermal fixed bed reactors, 162–183
  steam reforming, 111
Copper, low temperature shift catalysts, 83, 155
Cracking, see also Silica–alumina catalysts
  catalyst poisoning by organic bases, 11

Cracking—(contd.)
   fluid, moving and fixed bed reactors, 218
   fluid beds, 191
   n-hexadecane, 162
Cumene
   cracking, 10
   cracking over H-mordenite, 19

Damköhler number, modified, definition of, 169
Detoxification of poisoned catalysts, 8
Diffusion
   bulk, 33
   control in catalyst coking, 16
   counter diffusion flux ratio, 35
   deactivation and, 43–49
   effective, 36
   Knudsen, 33
   Knudsen in fixed beds, 150
   multicomponent, 37
   parallel pore model, 36, 37
   porous catalysts, 38
   random pore model, 37
   Stefan–Maxwell relations, 37
   surface, 33
   transition region, 34
Diffusional resistance, effect of fouling in fixed bed reactors, 175–179
Dispersion, loss of, 3
Dynamics of reactors
   effect of catalyst deactivation, 184–188
   hydrogenation of benzene, poisoned by thiophene, 188
   start-up effects in CO oxidation, 187

Effective thermal conductivity of catalyst pellets, 38
Effectiveness factor
   coking of single pellets, 123, 129, 131, 133, 137, 139
   definition of, 39, 40
   figures, 40
   independent poisoning, 92
   influence of external mass and heat transfer, 41, 42
   isothermal, influence of Thiele modulus, 40
   non-isothermal, 41, 42
   reversible poisoning, 102, 106

Electron microscopy,
   metal crystallite areas, 58

Ferric oxide/chromia
   high temperature shift catalysts, 83
   thermal sintering in reactors, 184
Finite thickness reaction zone, retracting core model, 225, 227, 228, 229, 233
Fischer–Tropsch catalysts, 5
Fouling, *see also* Coking, Metal deposition
   by coke deposition, 4, 110–140
   by deposition of feed stream impurity, 110
   parallel, 5
   parallel in catalyst pellets, 114
   parallel in reactors, 157–183
   series, 5
   series in catalyst pellets, 115
   series in reactors, 157–183
   transient analysis in single pellets, 138, 139

Grain model, *see* Particle-pellet model

Hydrogen sulphide removal by zinc oxide, 86
Hydrogenation
   benzene, non-isothermal poisoning in fixed beds, 156
   benzene, non-isothermal poisoning dynamics in fixed beds, 188
   ethylene, coking in fixed bed reactor, 161
   ethylene, pellet poisoning by oxygen, 104
Hydroisomerization of olefins, 160

Iron deposition on catalysts, 19

Langmuir–Hinshelwood mechanisms
   analysis in coking of single pellets, 117, 134, 138, 139
   coking, analysed by, 14
   fixed bed reactor, non-isothermal reversible poisoning, 157
   fouling in fixed bed reactors, 168, 171

Index

fouling reaction in fixed bed reactors, 175
reversible poisoning, 99
selective poisoning, 108
Lead deposition in automobile catalysts, 88

Macropores, 29, 30
Mass and heat transport, 31
  correlations, 32, 33
  external mass and heat transfer, 30, 31
  heat transfer coefficient, 32
  intraparticle heat transfer, 38
  intrapellet mass transfer, 30, 33
  mass transfer coefficient, 32
Maximum temperature rise in regeneration of coked pellets
  comparison of models, 231, 233
  effect of gas composition, 232
  effect of radiation losses, 234
  experimental results, 222, 223
Metal atom migration in sintering, 71
Metal deposition on catalysts, 19
  metal sulphide fouling, 4
  metal sulphide, pore plugging, 20
  pore plugging in single pellets, 140–143
  pressure drop due to, 140
Micropores, 29, 30

Naphtha purification for steam reforming, 86, 242
Nickel
  catalysts, regeneration from H_2S poisoning, 213
  catalysts, selective poisoning for acetylene hydrogenation, 108
  catalysts, simultaneous sulphur poisoning and coking, 183
  catalysts, sintering, 56
  catalysts, steam reforming, 22, 86
  supported catalyst, poisoned by thiophene, 105

Optimal operation of reactors
  optimal catalyst structures in pellets, 206
  optimal catalyst structures in reactors, 207
  start-up policies for deactivating reactors, 210
  steady-state conversion, 205
  time on stream, series reactions, 207

Particle migration, theory for sintering, 63, 65–69
Particle-pellet model, 226, 228, 229, 233
  fixed bed analysis, 256
Phosphorus, poisoning of automobile catalysts by, 88
Platinum catalysts
  control of automobile emissions, 87
  platinum/palladium catalyst, H_2S poisoning in fixed bed, 149
  platinum/rhodium gauze catalysts for ammonia oxidation, 84
  poisoning by thiophene, 49
  reforming, 12, 242
  reforming for isomerization of pentane, 118
  sintering, 60, 61
Poisoning
  analysis of irreversible, 88–97
  analysis of reversible, 97–103
  automobile catalysts, 87, 88
  bifunctional catalysts, 12
  correlations for, 50, 51
  deliberate, low temperature shift catalysts, 86
  guard reactors to minimize, 86
  low temperature shift catalysts, 83
  mass transfer control, 94
  metallic catalysts, 7
  metals, 8
  methanation catalysts, 84
  minimization of, 85–87
  non-isothermal effects, 103
  non-metallic catalysts, 9
  silver oxidation catalysts, 84
  steam reforming catalysts, 81
  toxic metals, 8
Poisoning, irreversible
  thiophene for benzene hydrogenation, 105
  thiophene for benzene hydrogenation in fixed beds, 156
Poisoning, reversible
  ammonia catalysts, caused by water vapour, 85

Poisoning, reversible—(contd.)
  ammonia catalysts, caused by carbon monoxide, 85
  copper/magnesia catalyst, caused by water vapour, 97
  ethylene hydrogenation, caused by oxygen, 104
  fixed beds, non-isothermal, 157
  hydrogenation catalysts, 85
Poisons
  arsenic, 83
  chlorine, 83
  sulphur, 82
Pore blocking, 13
Pore mouth deactivation, 45–49
  automobile catalysts, 88
Pores
  mean diameter, 20
  structural models, 35
  structure of, in catalysts, 29, 30
Profiles
  activity in coked fixed beds, variation with adsorption constants, 180–183
  activity in coked fixed beds, variation with Thiele modulus, 176, 178
  activity and concentration for poisoning, 91
  activity, concentration and temperature for fouling, 121, 125, 126, 130, 132, 136
  activity, concentration and temperature for poisoning, 94
  activity, fixed beds on coking, 166, 167
  coke concentration during regeneration, 216
  concentration, reversible poisoning, 100, 102
  concentration, temperature and activity in coked fixed bed reactors, 172, 174
  optimal temperature, 203
  poison wave, fixed beds, 149
  temperature, catalyst sintering in reactors, 184
  temperature, deactivating catalyst bed, 200
  temperature, fixed bed with catalyst ageing, 155
  temperature, fixed beds on coking, 165, 166
  temperature, non-unique steady states in fouled reactor, 185, 186
  temperature, regeneration using coke composition model, 241
  temperature, regeneration of coked pellets, 221, 222
  temperature, reversible poisoning, 104
  temperature, thermal sintering in reactors, 184

Reaction modulus, definition of, 169
Reactor dynamics, see Dynamics of reactors
Reactors
  coking of, 157
  coking, isothermal, 158
  coking, non-isothermal, 162
  dynamics with deactivating catalysts, 184–188
  heterogeneous and homogeneous models, 146
  Knudsen diffusion with poisoning, 150
  non-isothermal analysis, 154
  optimization, 190, 199
  plug flow with poisoning, 148
  poison adsorption in, 147, 148
  poisoning in, isothermal, 147
  sintering of, 183–186
Regeneration
  carbon burning rate, 126
  coke composition model, 235–242
  coked catalyst pellets, 219
  comparison of temperature profiles in pellets, 222, 223, 241
  composition of regeneration product gas, 215, 219, 244
  controlling steps, 215–217
  feasibility of, 212–214
  fixed beds, experimental results, 249–251
  fixed beds, models, 244–259
  fluidized bed catalysts, 218–219
  homogeneous model, 215, 220, 224, 228, 229
  isothermal models for catalyst pellets, 220
  isothermal two-stage models, 220

kinetics of, 214–218
mass balances for pellet models, 228
negative temperature step in feed, 259
perturbations, effect of, 256–259
reforming catalysts, 242
shell progressive model, 215, 220, 253
temperature runaway, 219, 244
Residua, metal deposition from, 19

Selectivity, 43
  effect of poisoning on, 107–108
  silver oxidation catalysts, 212
Separable and non-separable kinetics, 52
Sharp interface retracting core model, 224, 227, 228, 229, 232, 233
  inclusion of radiation heat losses, 234
Silica/alumina catalyst
  cracking of ethylene on, 17, 118
  cracking of hydrocarbons on, 18
  cracking of xylene on, 17
  oil cracking by, 118, 218
  poisoning of, 9
  poisoning of alcohol dehydrogenation by butylamine, 53
  poisoning by organic bases, 11
  sintering of, 3
Silver catalysts, selectivity and deactivation, 212
Sintering
  activation energies for, 61, 62
  atomic migration model, 69
  comparison of models, 72–74
  correlations for, 26
  experimental results on flat and microporous supports, 74–77
  inhibited growth model, 76
  interparticle transport, 64
  mechanisms for, 23, 63–74
  nickel catalysts, 62
  nickel on alumina catalysts, 23, 25
  oxides, 23

oxygen, influence of, 73, 75
palladium catalysts, 62
particle migration model, 65
platinum catalysts, 62
power law orders, 60
power rate laws for various mechanisms, 79
promotion by poisons, 26, 27
rhodium catalysts, 62
silver films, 26
Start-up, effect of deactivation, 187–188
Strength of catalysts, 6
Sulphur, catalytic removal from petroleum fractions, 19, 20
Supported metal catalysts, 12

Thiele modulus
  definition of, 39, 40
  effect on fouling in fixed bed reactors, 175–179
  modified, for deactivation processes, 44, 45
Tortuosity factor, 36
Trickle bed reactors for hydrotreating, 21

Vanadium
  deposition on catalysts, 19, 22, 110
  catalysts, for control of automobile emissions, 87, 88

X-rays
  diffraction of, for metal catalyst areas, 67
  low angle scattering for metal catalyst areas, 58

Zeolites
  fluid crackers using, 218
  mordenite, pore blocking by coke, 19
  poisoning of, 9
  quinoline titration of, 12